C000310870

What Readers A

From Chaos to Successful Distributed Agile Teams

"Anyone who manages or works on some sort of distributed team should read this book. It is a goldmine. Reading this book opened my eyes to a whole level of nuance and complexity I was missing."
—MIKE LOWERY, AGILE COACH

"Everyone should read the leadership chapter of this book, take it to heart, and then pass the book to your teams. Work together to become better, more agile teams, anywhere and everywhere."
—MICHAEL HERMAN, PRINCIPAL CONSULTANT, MICHAEL HERMAN ASSOCIATES.

"This isn't just a great book for distributed agile teams; it's a great book for any agile team"
—RYAN DORRELL, CHIEF SOLUTIONS OFFICER, AGILETHOUGHT

"Thanks to this book, I now understand our distributed team is actually a nebula team and I found a ton of tips that will help us improve our experimentation, communication, and collaboration. A practical book like this was long overdue."
—JURGEN APPELO, AUTHOR OF *MANAGEMENT 3.0* AND *MANAGING FOR HAPPINESS*

"A timely and practical book that is both pragmatic and compassionate—modern product development thinking in a context of healthy distributed teamwork. If you are an agile team member, leader, HR professional, coach, or virtual facilitator, this is your go-to guide for successful distributed teams."
—ELLEN GOTTESDIENER, PRODUCT COACH

"Remote (part-time or full) is a common reality. You now have a guide—packed with years of hands-on experiences and learnings—to help you understand and make your distribution situation work for you rather than against you. You'll discover the tips in this handy companion will serve your collocated and distributed teams for years to come. I wish I had this book 3-4 years ago—it would have saved me and teams I worked with much frustration and misunderstandings."

—MARCUS HAMMARBERG, AUTHOR OF AUTHOR OF
KANBAN IN ACTION AND SALVATION: THE BUNGSU STORY

"From Chaos to Successful Distributed Agile Teams is a tour de force—the best book on teamwork I've read this decade. Two days after starting the book I was implementing small experiments with my own distributed team. And, I'll be recommending this book to all of my clients."

—CHRISTOPHER AVERY PH.D., AUTHOR OF THE RESPONSIBILITY PROCESS:
UNLOCKING YOUR NATURAL ABILITY TO LIVE AND LEAD WITH POWER

"Distributed agile is not easy, but it is possible, and worth the journey. Their book emphasizes a people-centered approach to distributed agile—not just to enable a team to do its best work, but also to maintain connection, continuous experimentation, and learning."

—PILAR ORTI, DIRECTOR OF VIRTUAL NOT DISTANT

"Conventional agile wisdom assumes collocated teams. Yet, the reality of modern software development is that many teams have at least one, if not all, remote team members. You need to work with the people you have. This book offers pragmatic advice supported by real examples to support and nurture teamwork and high performance in remote teams."

—SHANE HASTIE, DIRECTOR OF AGILE LEARNING PROGRAMS, ICAGILE

From Chaos to Successful Distributed Agile Teams

Collaborate to Deliver

Johanna Rothman

Mark Kilby

From Chaos to Successful Distributed Agile Teams

Collaborate to Deliver

Johanna Rothman and Mark Kilby
Copyright © 2019 Johanna Rothman and Mark Kilby. All rights reserved.

Published by Practical Ink
www.jrothman.com

Practical **ink**

Cover art: © Csuzda | Dreamstime.com—Team Growth Photo
Cover: Lucky Bat Books

Ebook ISBN: 978-1-943487-10-3
Print ISBN: 978-1-943487-11-0

Contents

CHAPTER 9

Adapt Practices for Distributed Agile Teams 173

CHAPTER 10

Integrate New People Into Your Distributed Agile Team. 197

List of Figures

Acknowledgments

We thank these people who read and reviewed the book and provided feedback: Heidi Araya, Tonianne DeMaria, Jesse Fewell, George Dinwiddie, Ryan Dorrell, Dave Gordon, Mike Hansen, Shane Hastie, Michael Herman, David Horowitz, Sue Jasmin, Mike Lowery, Pilar Orti, Craig Smith.

We thank our editors, Rebecca Airmet and Nancy Groth. We thank Karen Billipp for the layout and Jean Jesensky for indexing the print book. Cover design by Brandon Swann.

From Johanna

I thank my clients and geographically distributed agile team workshop participants. You and my Pragmatic Manager readers and Managing Product Development readers helped me refine my ideas with your questions and comments.

I thank Shane Hastie for working with me to develop and lead several distributed team workshops and for writing our articles together.

From Mark

There are many who helped shape my thoughts in this book.

I have been fortunate to work with many distributed agile teams over the last 15 years who trusted me when we experimented together. You provided much of the inspiration for this book.

Through many conversations and collaborations (many online), I'm grateful for those who helped me refine my ideas: Jim Benson, Tonianne DeMaria, Michael Herman, David Horowitz, Pilar Orti, and Lisette Sutherland.

Much of our work in distributed agile teams arises from considering the art of the possible. I hope all of you join me in remembering Jean Tabaka for that inspiration.

Introduction

Distributed agile teams have a bad reputation: too often, they have problems starting and finishing the work. The managers don't know why. The team members don't know why. People wonder, "Why can't this team just get on with the work?" In the meantime, the team struggles to work as fast and as hard as they can.

For years, if you wanted guidance about how to be a geographically distributed agile or lean team, the answer was, "Don't do that."

"Stop being distributed" or "Don't use agile" is not useful advice. That would require sweeping changes in people's expertise, location, and the organization's ability to deliver products. That would result in disruption of work or possibly loss of valuable team members and product sales.

Distributed work is not the same as collocated work. But the agile principles can be adapted and applied to distributed teams. That's what this book is about.

We wrote this book for three audiences. First, for distributed and dispersed team members, so they can see how they might create communication channels and agile practices that work.

Second, we address those who facilitate and serve distributed teams. We've seen a variety of possible team "leaders." These servant leaders might be coaches, agile project managers, or Scrum Masters. Sometimes, these servant leaders are technical leaders or technical managers. Whatever their title, they facilitate and serve the distributed team.

Our third audience is the managers, executives, and organizational coaches/facilitators who want to take advantage of global talent *and* agile approaches for frequent delivery of value to customers. These organizational leaders create the environment—the culture for collaboration—in which the teams then work and evolve.

We assume that all of these people—team members, team leaders, and organizational leaders—want everyone on the distributed or dispersed team to work to the best of their capability. However, too often, distributed and dispersed work frustrates everyone. That's because too many teams retain their collocated mindset for their distributed or dispersed teams.

We've seen three necessary mindset changes for successful distributed agile teams. The first mindset change is the agile mindset of encouraging and managing for change. When the team encourages experimentation for *everything*, the team manages how and when it decides to change.

The second mindset change when moving to distributed agile teams is the emphasis on communication and collaboration. When the team creates communication and collaboration norms for *everyone*, the team can eliminate many of their impediments to delivering value.

The third mindset is to use agile principles—not common practices— to create a distributed agile team. When teams use principles to create *their* practices, they adapt their agile approach to fit their context.

With this mindset of experimentation, communication and collaboration, and using principles over practices, distributed agile teams can succeed. Without that mindset, teams work too "slowly" and everyone—from the team members to the executives—becomes frustrated.

That's when people say, "Agile doesn't work for distributed or dispersed teams. It doesn't work for us."

You can create high-functioning, high-performance geographically distributed agile teams. Your teams might change—you may decide that the current team makeup doesn't fit anyone's needs. You might decide to recreate teams with more hours of overlap. But, you can succeed with geographically distributed agile teams.

Agile and lean approaches will make your problems transparent. Because they do, you may decide that agile geographically distributed teams reveal other problems in your organization. With this transparency, you can make better decisions.

We assume you, our readers, are somewhat familiar with many of the agile terms and practices. We are not going to explain them all in this book. Instead, we offer references to other books you might want to read to gain deeper understanding. We *will* explain how distributed agile teams may adapt specific practices to be successful.

As you read the book, you might notice we use words such as, "We have found . . ." That phrasing refers to our combined 55+ years of experience with distributed and dispersed teams. We have worked in distributed and dispersed teams in various roles: developer, tester, project manager, program manager, manager, consultant, coach, and workshop leader. Your experience might be different from ours.

Let's start.

CHAPTER 1

Distributed Agile Teams Are Here to Stay

Many pundits, via podcasts, articles, and books, have declared remote work the wave of the future.

If you've always commuted to the office and worked with other people in person, this idea of remote work might seem strange. You might even think, "This can't possibly work." How can you learn about your colleagues? How can you collaborate? If you work from home, how can you structure your day?

Agile approaches can answer these questions for geographically distributed teams.

Back when the signatories of the Agile Manifesto released those values and principles, we had insufficient technology to manage remote communication and collaboration. Now, technology allows us to connect and collaborate when we are not face-to-face.

Agile principles amplify the need to connect and collaborate on a frequent basis to deliver value. So while a person may be "remote" from their colleagues, they can now be very connected with those same colleagues through a number of rich and natural communication channels. More importantly, they have even more flexibility than in a traditional office environment to decide when to engage in this intense collaboration with their remote colleagues.

However, using "standard" agile practices does not guarantee success for distributed teams. To make distributed agile teams work, everyone shifts their mindset to a culture of experimentation, communication and collaboration, and principles over practices.

1.1 Understand Agile Teams

We use the word "team" in this book. Based on the work of Katzenbach KAT99 and our experience, an agile team is a cross-functional group of people who:

- Have the necessary skills and capabilities their team requires to deliver on their objectives
- Are committed to a common purpose or goal
- Are interdependent and therefore make commitments about the work to each other
- Learn to understand each other: their strengths, weaknesses, and preferences
- Plan and deliver the work in a collaborative fashion, which can include co-designing, co-creating, pair reviewing, or mobbing on the work
- Reflect together, reviewing their work and their process in a collaborative fashion
- Are committed to one team and one team only.

Often, collocated team members depend on physical connection. They can work face-to-face. They can mob with the entire team in one room. They can go to lunch as a team to learn more about each other.

A distributed or dispersed team has team members apart from each other. No one has a physical connection to all the other members of the team. However, for successful distributed agile teams to work, each team member must build social relationships with each other member.

Collocated, Distributed, or Dispersed?

What is "close enough" for collocation? It's close enough for easy collaboration. The *optimal* distance for communication frequency is less than eight meters. Once team members are separated by 30 meters—and this includes separation by stairs or elevators—their chance of off-the-cuff communication declines dramatically. If your team members are not within 30 meters of each other, you have a distributed team of some type.

Distributed teams have people in several locations, too far away to be collocated with each other. Some people might be close enough to actually walk to each other. "Several" locations can range from two (most people are collocated and one or two are remote, which we we refer to as a satellite team) up to the the number of team members minus one. When at least two people are collocated with each other in multiple places, we refer to it as a cluster team.

Dispersed teams have all people remote from each other. No one is close enough to walk to see each other. The entire team works virtually. We refer to this as a nebula team because too many people confuse dispersed with distributed.

We'll talk more about the kind of team you have and the characteristics of the team types in *Identify Your Distributed Agile Team Type* on page 65.

 You might not realize that a team with all members on the same campus can be a distributed or dispersed team. However, if floors or buildings separate your people by at least 30 meters, the team is distributed.

However, agile approaches still may not be for your organization or team. So let us explore why you might want to work in a distributed team, why agile principles may be an important aspect of that work, and how you can ask some critical questions to determine if this is the right type of work for you, your teams, and your organization.

1.2 Why Distributed Teams?

Why do you want to use distributed teams?

We know of several organizations that are completely distributed or dispersed by choice. We suspect there are more organizations who would like to be fully dispersed.

We've seen at least three good reasons to create distributed teams:

- Companies want the ability to hire talented people anywhere in the world or retain people who move.
- Companies need a resilient workforce that can continue despite road closures, bad weather, or other physical challenges in commuting to an office.
- People want the ability to avoid commute time (and possibly energy cost) for their personal benefit.

You may have other reasons for your distributed team.

We've also seen reasons that don't satisfy the organization's needs. Here are several:

- A leader believes low salaries will save project money (see *Trap: Save Money with Lower Salaries*, page 18).
- A leader wants to support colleagues in another country. See *Trap: We Can Hire Experts Anywhere* on page 221.
- A leader thinks about people as resources instead of as people, and they think they can split the work to make people more efficient. See *Think in Flow Efficiency* on page 10.

Once you clarify why you need a distributed team, next consider the decision to use agile approaches for distributed and dispersed agile teams.

1.3 Agile Approaches Focus Distributed Teams

Agile approaches can help distributed teams and team members respond to change. For example, while writing this book, we encountered numerous hurricanes and snowstorms. If we had only been able to work as a collocated pair, we would have had to move to be close to one another *and* we would have lost days of work because of the weather. Instead, we were able to adapt our work and create a resilient project because we worked as a distributed agile team.

We applied the three mindsets necessary for successful agile teams that we mentioned in the Introduction:

1. Manage for change. This means experiment.

We experimented with everything on our project—from writing approaches, tool selection, writing hours, gathering reviewer feedback—and more as we worked through this project. We didn't select one approach for our project. We used the idea of small safe-to-fail experiments with double-loop learning to help us progress through the work.

You might discover that experimentation also applies to your agile approach for your team.

2. Emphasize communication and collaboration.

We selected tools that we could both use that were readily available, inexpensive, and met our security needs. (See Appendix A for *Our Toolset* on page 253, which might not be your toolset.)

For your team, make sure that everyone has access to readily available, inexpensive, and secure technology that allows distributed teams to collaborate. Make sure the tools aren't a barrier to collaboration.

3. Use agile principles, not practices.

We see too many teams assume that focus on a particular practice, such as a standup or backlog refinement, will translate directly to a distributed or dispersed team. Practices don't always translate directly. In this book, we'll define and discuss eight agile principles.

1. Establish acceptable hours of overlap.
2. Create transparency at all levels.
3. Create a culture of continuous improvement with experiments.
4. Practice pervasive communication at all levels.
5. Assume good intention.

6. Create a project rhythm.
7. Create a culture of resilience.
8. Default to collaborative work.

Distributed and dispersed agile teams are here to stay. Distributed team members can experiment, communicate, and collaborate. Distributed agile approaches can help team members and projects become more resilient. And, everyone can benefit if people are not tied to a corporate office.

Questions to Ask First

Before you make the decision to use agile approaches for your distributed team, ask these questions:

Are we willing to let our team experiment with different practices over short durations to see what works best for everyone on the team?

Are we willing to default to collaboration over solo work so the team can collectively find the best solutions?

Are we willing to challenge what's in the product and how it works, *and* challenge our assumptions of how the team works, all to discover and deliver a successful product?

If the answers to any of these questions are "No," don't despair. Please do continue with the principles in this chapter and the next, and then read *When Agile Approaches Are Not Right For You* on page 17.

Now, let's examine the three mindsets in more detail.

1.4 Create a Culture of Experimentation

The first mindset shift is to create a culture of experimentation.

Collocated agile teams tend to experiment with practices. These teams have many resources available in the form of books, courses, and tools to support their experiments. Distributed agile teams—and their leaders—often need to experiment even more. Distributed and dispersed teams now have a variety of technologies to support their experimentation, even though there might not be books or courses to guide experiments.

Consider this story of a team who learned how to create small stories (agile requirements) and deliver those stories every day.

One Team Learned to Experiment as They Worked

The Search team is distributed across three time zones. Stefan, a developer, lives in Romania (UTC+2) . He works with Pierre (the tester) and Madeleine (the product owner) in Paris (UTC +1), and Tom, another developer, in Southampton, UK (UTC).

Stefan and Pierre have extensive experience creating and testing databases with an eye towards performance. They have access to other people in the company—New York (East Coast) and China. However, their team works across those three time zones in Europe.

Stefan, Pierre, and Madeleine learned about agile approaches by first reading books, and then by traveling to one city to take public agile classes together. As an entire team, they decided to try their own agile approach. They created a kanban board to see where the work was and their flow of work. They soon learned that Pierre was overloaded and that Madeleine was too optimistic about what the team could deliver and how fast.

The team created hypotheses: if they reduced their total Work in Progress (WIP), could Pierre keep up with testing? If Stefan and Pierre tested differently, would that make a difference for Pierre and Madeleine? If Madeleine created smaller stories, would the team and the customers have better outcomes?

> As the team worked together on their hypotheses, they learned to create shorter feedback loops inside the team and with customers. They created another hypothesis: what if the entire team spoke with customers every so often to learn more about what the customers needed?
>
> Madeleine was concerned at first, because she wanted to make sure the team members weren't insensitive to customers' needs. However, once the team members assured her they would be respectful, the entire team participated as a team with internal customers in the European time zones. They had sufficient hours of overlap.
>
> This team has since worked together for a couple of years choosing other experiments with short feedback loops, and they have a fairly dependable throughput. They finish at least one story a day. They have great relationships with their European customers. The team doesn't yet know how to work well with the people across more time zones, but they are working on that.

The Search team embraced experimentation with everything they do.

Experiments can come from the leadership or from the team itself. We'll talk about this more when we discuss the agile principles.

 Running too many experiments at once can lead to "change fatigue." Instead, run experiments with a clear hypothesis and a short duration.

Instead of an "all-in" total change to using one specific agile framework, consider trying small experiments to apply agile principles in your teams. We have found that using agile approaches in existing distributed teams has actually improved the work, the distributed working environment, and the morale of team members.

1.5 Shift to a Mindset of Collaboration

The second mindset shift is toward collaboration and communication inside the team. Many distributed and dispersed teams find this shift the most difficult to build and maintain.

In fact, one of the most important elements of successful agile teams is having a sufficient number of hours of overlap. Without sufficient overlap, communication becomes asynchronous instead of synchronous, and collaboration becomes very difficult.

When a "Team" Isn't a Team

The Payments team belongs to the same organization as the Search team. Payments is dispersed across many time zones: China (UTC +8), Sydney (UTC +10), London (UTC +0), Boston (UTC -5), and San Jose (UTC -8).

The Payments "team" doesn't feel like much of a team. They do not have sufficient hours of overlap to collaborate. However, they have the will to experiment. Here is what they tried over two years:

- They got together for two weeks to learn how to work together.

- They used a kanban board with a planning and reflection cadence to see their workflow.

- They tried to create very small stories and use handoffs, like those described later in *Follow the Sun* on page 44 to finish work.

They had marginal success as a team. In addition, they all felt pulled by their functional managers to do other work. They had trouble affiliating as a team.

As a result, the Payments team had almost 100% turnover. When the fourth person left the team, he wrote a memo to the managers. He explained that instead of asking people across many time zones to

---

work together, they could either create several Payments teams who had more hours of overlap or create Payments teams in a couple of offices.

That memo helped the managers realize they had to stop thinking about people as cogs and start thinking of people as humans.

The managers decided to create an East Payments team and a West Payments team. The East team refers to the UTC + time zones and the West team refers to the UTC - time zones. By dividing one "team" with too few hours of overlap into two real teams who can collaborate, the teams were able to finish their feature sets in reasonable time frames.

The experience of the Payments team is what gives distributed and dispersed teams—agile team or not—a bad name. This can change when teams consider and apply the agile principles.

1.5.1 *Think in Flow Efficiency*

How we think of teams is a part of that second mindset shift to communication and collaboration.

Too often, managers think of the individuals on a team, not the team itself. When managers think of people as "resources," managers think they or the teams can split work to make people more efficient. We split this work by work type, creating experts. The name for this is Resource Efficiency.

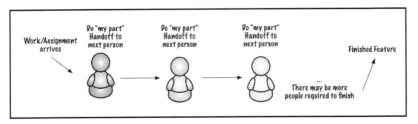

Figure 1.1: Resource Efficiency

Figure 1.2: Flow Efficiency

The idea of flow efficiency comes from *This is Lean*, MOA13. When a team works as a team—not individuals—they raise their team-based throughput over what any one person can contribute alone. The name for this is Flow Efficiency.

Successful agile teams implicitly use flow efficiency. The entire team collaborates to move the work to done. The team doesn't suffer delays from handoffs. And, information doesn't become stale.

Successful *distributed* agile teams reinforce the idea of flow efficiency. The more the team works together, the faster they are. The team learns how to work together, how to learn together, and how to solve problems together. They have few delays in their team process.

That means that the cost models of moving some work to a less expensive location doesn't work if the team has insufficient hours of overlap.

A complete feature team can work anywhere in the world, assuming all the people on that team have sufficient hours of overlap. See *Measure Project Costs* on page 229 for a more thorough discussion of how to keep project duration short and project costs low. As long as that team is independent, their wage cost may well be lower than other teams' wage costs.

However, if teams or people who have to collaborate together have fewer than four hours of overlap, the delays between the people or the teams may well overcome the wage cost savings.[1]

---

[1] https://www.jrothman.com/mpd/agile/2010/03/wage-cost-and-project-labor-cost/

If you serve a team with people who have insufficient hours of overlap, and your managers don't realize that they are creating a Cost of Delay (see *Measure Costs of Delay in Distributed Teams* on page 228) with the people, map the value stream and measure the team's cycle time.

It's possible to save money on wages with full feature teams. As long as those teams don't have interdependencies on other teams, their throughput will be high enough to compensate for any starting delays.

1.5.2 Non-Collocated Teams Deserve Face-to-Face Time

When we think in flow efficiency, we optimize for the team's deliveries, not a person's. That means non-collocated teams might need to meet on a regular basis.

People often ask us how frequently the non-collocated teams should meet in person.

We've seen teams discover better collaboration and more throughput when they meet for a minimum of a week once a quarter.

If you're not sure how often your team should meet in person, consider asking the team to *Map the Value Stream to Visualize Cycle Time* on page 46 and to measure their satisfaction. When teams experience an increase in cycle time and a decrease in personal satisfaction, they tend to have systemic problems. People often need to meet to discover and fix systemic problems.

If you are starting or restarting a non-collocated team, the cheapest thing you can do is to bring the team together for a week or two.

In Week 1, create a temporary team room so people can work together, physically, in one place. The team can build mutual understanding and trust as they work together. Take this time to understand each other's strengths, work preferences, and typical work hours. (See *Build Respect with Working Agreements* on page 147.) When the team works together, they can explore possibilities for: more hours of overlap; how to plan; how to hand off work; how to offer feedback to each other; and how to reject work that does not meet the team's standards.

If the team has a Week 2, consider conducting the second week in the team's virtual workspace, even though the team remains physically together. When teams practice in their virtual workspace, they select their tools for planning, development, review, and other possible tasks. For risk management, the team might also select and practice with their secondary collaboration tools and backchannels. In this second week, the team agrees on, practices with, and manages the risks of their virtual workspace. Because the team is collocated, they can also choose to conduct a quick retrospective each day to refine how they work together. We recommend the team transition to using their virtual workspace for their retrospectives during this week. This second week creates and cements the team's working agreements.

We've seen teams learn to experiment in a one day face-to-face, when they could not understand how to experiment at all when they were dispersed. After that one day experiment, they brainstormed three ways to experiment back in their offices. They continue to experiment.

The payoff for face-to-face time might be when:

- Teams charter their project in an hour or less, and then are able to discuss what's in and out of the release with the product owner.
- Teams learn how to pair, swarm, and mob when they are in the same room and they are then able to decide how to pair, swarm, and mob when they are back in their offices.
- Teams discuss their interpersonal challenges of how to offer and receive feedback and coaching from each other. Being face-to-face allows them to learn how to talk with others and how to make sure the other person heard them.

You might think the travel expenses are "too high." However, the team saves time and money through faster clarification of the vision, mission, goals, and working agreements. That clarification speed is worth the investment.

If you have a large team, those costs can appear prohibitive.

However, capitalizing software can offset the operational expenses. If you bring a team together for just one week, and they release a feature that you can capitalize, you might "make" all your money back in just one week. And, the team now has the skills to continue to release finished product more often.

If your team cannot collocate for at least a week, consider visualizing the team's value stream and measuring the team's *Measure Costs of Delay in Distributed Teams* on page 228 to visualize and understand your team's costs when you can't bring people together.

Now that we've addressed the experimentation and collaboration mindset, let's set the stage for principles over practices.

1.6 Review the Agile and Lean Principles

We assume you are familiar with the four values of the Agile Manifesto. However, you might not be familiar with the twelve principles of agile software development.[2] This list is a paraphrase of the principles. We expect any agile team (distributed or not) to use and live by these principles.

1. Deliver early and often to satisfy the customer.
2. Welcome changing requirements.
3. Deliver working software frequently.
4. Business people and developers must work together.
5. Trust motivated people to do their jobs.
6. Face-to-face conversation is the most efficient and effective method of conveying information.
7. Working software is the primary measure of progress.
8. Maintain a sustainable pace.
9. Continuous attention to technical excellence and good design enhances agility.

---

[2] http://www.agilemanifesto.org/principles.html

10. Simplicity—the art of maximizing the amount of work not done—is essential.
11. The best architectures, requirements, and designs emerge from self-organizing teams.
12. Reflect and adjust at regular intervals.

Changes in technology in the last decade have redefined what "face-to-face" can mean for distributed teams.

In addition, we find the two pillars of lean[3] especially helpful for distributed teams:

1. Respect for people
2. Continuous improvement

And the principles of lean software development as in *Lean Software Development: An Agile Toolkit*, POP03 provide a useful perspective when we consider how to support distributed teams:

1. Eliminate waste
2. Amplify learning
3. Decide as late as possible
4. Deliver as fast as possible
5. Empower the team
6. Build integrity in
7. See the whole

Agile teams tend to create respect for people inside the team as a matter of course. In addition, we have entire chapters about safety and respect: *Create Your Collaborative Team Workspace* on page 111 and *Build Respect with Working Agreements* on page 147.

1.7 Is a Distributed Agile Approach Right for You?

Not every organization can use a distributed agile approach. A distributed agile approach might not work for your organization if:

---

[3] http://www.leanprimer.com/downloads/lean_primer.pdf

- Your organization wants to use a hierarchy instead of a networked approach to the work. Distributed team members need to be able to solve their problems as a team without relying on the management hierarchy.
- Your organization only tracks expenses instead of value.
- Your organization measures people on their resource efficiency, instead of a team-based flow efficiency.
- Your organization prefers to establish core hours for all employees instead of letting teams decide their optimal core hours for working collaboratively.
- Your organization prefers to center all planning and activity around "headquarters" instead of teams that may be located around the globe.
- Your managers believe that people only need to know certain things, not the entire context of a project or some work. (e.g., stakeholders, key business goals, return on investment)
- A management-designated leader—some single person—needs to be involved with every team decision, instead of letting the team decide for themselves how to work. Sometimes, that leader is actually a small cluster of people, often close to headquarters.
- The team members prefer to be told what to do, when to do it, and how to do it, instead of taking work and determining who they need on the project and how to finish it.
- Team members prefer to keep to themselves instead of sharing some personal context and goals so teammates understand where their teammates' strengths are and what motivates their best work.

Agile approaches are much more than a project change. Agile approaches change the organization's culture.

We have found that people, teams, and managers who believe in and use the agile and lean principles are able to move past behaviors and beliefs like those listed above so that they can use an agile approach. However, sometimes, an agile approach isn't right for a team.

1.8 When Agile Approaches Are Not Right For You

Adopting agile practices without thinking about the principles does not make sense for many distributed teams. (See *Adapt Practices for Distributed Agile Teams* on page 173.)

Consider whether agile approaches make sense for your teams. For instance, some teams have too few hours of overlap to take advantage of agile approaches. In that case, consider asking them to use a different project life cycle. Teams have many choices other than only waterfall or agile approaches.

Teams can use iterative and incremental approaches to the work, without requiring the collaboration and learning from an agile approach. See *Manage It! Your Guide to Modern, Pragmatic Project Management*, ROT07 and *Agile and Lean Program Management*, ROT16A for details about choices. In addition, see *What Lifecycle? Selecting the Right Model for Your Project.*[4]

If you decide an agile approach is not right for the team, do consider organizing as a collaborative cross-functional team. Also, ask the team how they might limit their work in progress (WIP), and how they will know they are done. Teams who work via the agile principles, regardless of the project life cycle, often have better outcomes than teams that don't. Even if they don't use a recognizable agile approach.

Even if the teams might not look like they're using an agile approach, you, as a leader in the organization, can use an agile mindset to create the culture in which distributed teams can flourish.

1.9 See Traps That Prevent Successful Distributed Agile Teams

We've seen the future of successful distributed agile teams. We've also seen too many traps—often at the management or executive levels—that prevent successful distributed agile teams.

---

[4] https://www.jrothman.com/articles_/2008/01/what-lifecycle-selecting-the-right-model-for-your-project/

We've seen these traps prevent success in distributed agile teams:

- Assuming that lower salaries will create a lower project cost.
- People have an insufficient experimentation mindset.
- The teams were not originally created to use a culture of collaboration.
- Imposition of a specific agile approach and practices, rather than using principles for this team's context.
- Managers hold traditional expectations when distributed teams move to agile.

1.9.1 *Trap: Save Money with Lower Salaries*

For too long, managers have believed that software teams (or other knowledge work) is similar to factory work. Factory work often uses a divide-and-conquer approach to dividing up work to see it finish.

However, knowledge work includes innovation and learning as key parts of what a team does. Team members collaborate to discover the product as they build it.

When you separate team members by a few hours of overlap, the team has trouble working together as in *Think in Flow Efficiency* on page 10.

See *Avoid Chaos with Insufficient Hours of Overlap* on page 41 for the details.

1.9.2 *Trap: Insufficient Experimentation Mindset*

Agile approaches are based on collaboration and experimentation because the team receives frequent feedback. Distributed teams need to experiment even more than collocated teams, because they don't have the easy opportunity for off-the-cuff feedback.

Managers, especially, need to build their experimentation mindset, so they can consider alternatives for the product and their processes. When leaders experiment, they empower the teams to do so.

Consider these options:

- Ask the question, "What are our hypotheses, what options do we have to test them, and how will we measure the results?" for everything: especially the team's practices and the product features.
- If the team is new to experimentation, set a timebox and ask them to choose the option where they can most easily and regularly gather data. Start with the smallest possible experiment.
- *Map the Value Stream to Visualize Cycle Time* on page 46 for the team on a regular basis to understand the team's work times and wait times. Use that data to visualize potential experiments.

Small frequent experiments help distributed teams become more effective, regardless of their agile approach.

See *Focus on Principles For Your Distributed Agile Teams* on page 23 to understand how to think about successful distributed agile teams. In addition, read these chapters:

- *Identify Your Distributed Agile Team Type* on page 65
- *Communicate to Collaborate* on page 87
- *Create Your Collaborative Team Workspace* on page 111
- *Cultivate Your Distributed Team's Agile Culture* on page 129
- *Build Respect with Working Agreements* on page 147
- *Adapt Practices for Distributed Agile Teams* on page 173
- *Integrate New People Into Your Distributed Agile Team* on page 197

1.9.3 Trap: Teams Were Not Originally Created for Collaboration

We described how to *Think in Flow Efficiency* on page 10, which might be a new idea for you.

The more everyone understands flow efficiency, the more they will foster collaboration in the teams. If your team was originally created for handoffs, rethink the flow of work through your team. See *Map the Value Stream to Visualize Cycle Time* on page 46.

In addition, read these chapters:

- *Create Your Collaborative Team Workspace* on page 111
- *Cultivate Your Distributed Team's Agile Culture* on page 129
- *Build Respect with Working Agreements* on page 147
- *Integrate New People Into Your Distributed Agile Team* on page 197

1.9.4 *Trap: Imposition of a Specific Agile Approach*

Each team needs its own rhythm, which we'll discuss in more detail in *Focus on Principles For Your Distributed Agile Teams* on page 23. Instead of focusing on specific practices, ask the teams to use the *Agile and Lean Principles* on page 14 and learn to release value as often as possible. Teams might deliver at varying cadences—and the key is to keep those cadences as short as possible.

In addition, please see these chapters:

- *Create Your Collaborative Team Workspace* on page 111
- *Cultivate Your Distributed Team's Agile Culture* on page 129
- *Build Respect with Working Agreements* on page 147
- *Adapt Practices for Distributed Agile Teams* on page 173

Too many people believe specific agile approaches can work for any team. We have seen successful agile teams use the principles to adapt practices.

1.9.5 *Trap: Managers Hold Traditional Mindsets*

Managers hold the culture that allow distributed agile teams to succeed.

We described the three mindset shifts to: experimentation, collaboration, and principle-based agile approaches. To create a successful distributed agile culture, managers need to change their mindset, also. This might challenge the best managers, because *their* managers haven't changed the management measures.

Please do read:

- *Focus on Principles for Your Distributed Agile Teams* on page 23
- *Cultivate Your Distributed Team's Agile Culture* on page 129

Agile approaches change the culture, not just of the agile team, but for the organization. Successful agile managers work up and down the hierarchy, helping their colleagues and *their* managers change the measurements and desired outcomes. See *Lead Your Distributed Agile Teams to Success* on page 225.

1.10 Now Try This

1. Write down why you or your organization wants to use (or is using) distributed teams. If one of the reasons involves cost, review the traps in this chapter. Now, define your business reasons for a distributed agile approach.
2. Consider how you, your team(s), and organization might use experiments. Does everyone on the team have an equal voice in proposing possible experiments about product and process?
3. Consider how you, your team(s), and organization currently collaborate. Is it possible for teams to use flow efficiency, rather than resource efficiency, to collaborate on the entire project, from defining the requirements through to delivery?

Next, let's see the principles that successful distributed agile teams use to create great working team environments.

CHAPTER 2

Focus on Principles to Support Your Distributed Agile Teams

Successful distributed agile teams control their context.

When teams can't control their context, they cannot easily deliver. They have too many delays. They don't communicate or collaborate well.

Instead of specific practices, consider these principles to create your team's context, your system of work. These principles make it possible for teams to use agile approaches.

1. Establish acceptable hours of overlap.
2. Create transparency at all levels.
3. Create a culture of continuous improvement with experiments.
4. Practice pervasive communication at all levels.
5. Assume good intention.
6. Create a project rhythm.
7. Create a culture of resilience.
8. Default to collaborative work.

When the organization's culture and the teams align with these principles, they create and refine a workable distributed agile culture. Misalignment with these principles can lead to misalignment with the goals of the business, the work of the teams, and can even be destructive to personal lives.

Let's start with the most important principle for a successful distributed agile team, that of hours of overlap.

2.1 **Establish Acceptable Hours of Overlap**

Teams can't effectively collaborate or communicate when they don't share enough working hours.

We have found that teams with less than four hours overlap have trouble adhering to any of the agile principles, never mind producing a quality product in a reasonable amount of time.

However, when teams have at minimum a half-day of overlap, they tend to find ways to work in flow as teams, not just as individuals. When teams find a way to work in team flow, they finish better work faster.

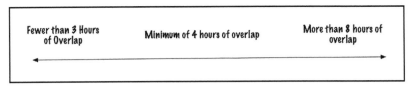

Figure 2.1: Hours of Overlap Principle

Most teams don't improve with more than 8 hours of overlap. This occurs when team members have trouble stopping work. When people don't stop working, they tend to burn out. They make mistakes. They are not at their best.

Conversely, when team members have fewer than three hours of overlap, they start work too early, stay too late, or otherwise interrupt their lives for work. We know of people who regularly wake up at 3 A.M. for a conference call, return to sleep, wake up at 6 A.M. for their day, and have another call at 10 P.M. We consider this behavior insane. These people have the same problems as people who work too much: they burn out, make mistakes, and are not at their best.

Too few hours of overlap give distributed agile teams a bad name.

Every team has its own acceptable hours of overlap. Maybe your team needs six hours of overlap to produce a high quality product. We recommend you default to at least four hours of overlap. Consider creating new teams with as many hours of overlap as possible. If you are having difficulties with the other principles in this book, return to this principle first.

We'll talk about some options and ideas for when your team has few or no hours of overlap in *Avoid Chaos with Insufficient Hours of Overlap* on page 41. However, most teams with zero hours of overlap cannot use an agile approach. See *When Agile Approaches Are Not Right for You* on page 17.

2.2 Create Transparency at All Levels

Transparency invites collaboration and creates meaning.

You might not be able to share everything, such as financial information. However, consider how much transparency you can create for the organization's strategy, the project portfolio, the why behind the products, the work that everyone does.

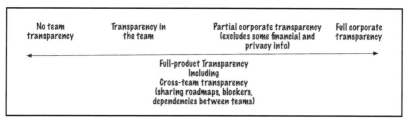

Figure 2.2: Create Transparency Principle

If we focus on the business of the team, product development, the sweet spot is "Full product transparency." If we focus on business agility—the ability of the organization to be fully agile—the sweet spot is full corporate transparency. This book is about product development in distributed teams, which is why "Full product transparency" (including cross-team transparency) is the sweet spot here.

Who needs to see what in your company? Agile teams own their features and their process. The features solve problems for specific customers in a potential time frame. When teams have full product transparency, they can see everything they need to see: who the customers are; how the product and new features solve specific customer problems; and any other goals for this work.

In the case of full corporate transparency, team members become partners in the business and therefore need complete transparency.

Siloed teams often have very little transparency. Once team members understand the product context and know what other team members are doing, they have "Transparency in the team." However, one team may not have complete transparency into the work of other teams working on the same or related products in a portfolio. This is typical of the new agile team that begins to master agile practices within the team but not how to carry agile principles outside the team yet to a program (a collection of projects with one business objective).

"Full product/cross-team transparency" occurs when all the teams have sufficient context to collaborate and deliver toward the product goals. They not only understand their product context and the context of other teams but also understand the product roadmap for discovery and delivery.

Teams with some collocated members might have one-on-one discussions and then explain their conclusions in a dedicated team backchannel.

Teams with sufficient hours of overlap may discuss the roadmap in real time with the product owners and possibly the product value team.

Teams with few to no hours of overlap have trouble creating and maintaining this full product context. That's because agile approaches allow and encourage us to change the roadmap and the ideas for the future (if not the current iteration's backlog). The product owner (PO) often needs to change the roadmap. Unless the PO uses a pervasive method of communication, like an all-group or all-team backchannel, some team members won't be aware of or understand the changes.

2.3 Cultivate Continuous Improvement with Experiments

Experiments allow teams to discover their preferences, figure out what success means, and learn and improve continuously. Consider how you can improve yourself first—and then improve your team. This goes for team members and leaders.

We have discovered that challenging ourselves first proves to our teams that we are invested in improvement. Selecting experiments for yourself will help others to take some chances in learning and trying new things. Those experiments help the team realize they can practice continuous improvement.

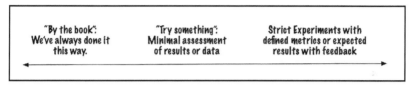

| "By the book": We've always done it this way. | "Try something": Minimal assessment of results or data | Strict Experiments with defined metrics or expected results with feedback |

Figure 2.3: Continuous Improvement with Experiments Principle

Teams require a safe environment in order to experiment. They also have to be owners of their process. Resilient teams are also better at experimenting. This is what we mean by considering the agile principles instead of the specific practices.

So where might a team begin with continuous improvement? Especially when a team transitions from a more traditional culture to an agile culture, many people say things such as, "We've always done it this way."

Agile approaches encourage double-loop learning, where we specifically challenge our assumptions about everything: the product we develop and the process by which we develop it.

We have seen that when individuals are open and honest about their personal experiments such as, "I'm going to ask for help if I'm stuck for 15 minutes," the team tends to be more open to team-based experimentation.

That first experiment might be the most difficult. However, we have found that in distributed agile teams, experimenting helps teams discover their best ways of working—for now. Those ways might change over time, so encourage your team to continually experiment.

2.4 Practice Pervasive Communication at All Levels

Communication is critical to distributed teams. Because they are not collocated, they have few, if any, opportunities to interact casually with other team members.

Distributed teams need several levels of communication. First, is the "why" for this project. Teams need to build shared understanding of the purpose of their work: Who is the work for and what is the value those people will receive?

Once the team knows why they should do the work, the entire team can collaborate on their plan for the work. (See the section *Plan as a Team* on page 183 for more information.)

Your distributed team might need access to the stakeholders. We've seen situations where—because everyone is distributed—the stakeholders are more ready to collaborate with the team. When the team is distributed, the team (and the stakeholders) often already understand how to show each other the product in progress and the customer/stakeholder environment. Successful distributed teams learn to discover ways to interact with stakeholders to receive their feedback on product progress.

Finally, the team defines how they will do the work—the how for the work and the how for communication. Successful distributed agile teams often have many channels of communication, even in meetings.

Figure 2.4: Pervasive Communication Principle

Avoid Over-Communication

One organization has improvement days called hackathons. Anyone can do anything to improve the work or the products as long as they share the results of that work.

When they first introduced these days, the VP announced the hackathon at a face-to-face gathering. The coaches facilitated but everyone drove their own work. Coaches used all the communication channels as reminders for the first few hackathons. After the first couple of hackathons, the coaches did a survey to see what people would change. The feedback was overwhelming:

> Over-communication! Please announce in our hackathon chat channel ONLY.

The hackathons came at a regular cadence and everyone in the organization wanted minimal disruption to their normal work, so they felt this one reminder was sufficient.

You can't know the right level of communication ahead of time. Create feedback loops and adjust as quickly as you can.

2.5 Assume Good Intention

When we assume good intention, we are congruent, thinking of ourselves, the other person, and the situation or context. See the Satir Congruence model discussion in *Software Quality Management, Vol 2: First-Order Measurement,* WEI93.

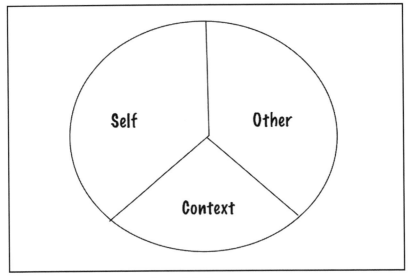

Figure 2.5: Congruence

When we are congruent, we think and act in balance: what we want and need, what the other person wants and needs, and what the situation needs. We might be incongruent in several ways: if we ignore the other person, we blame that person. If we ignore ourselves, we placate the other. If we ignore both of us, we are super reasonable. If we ignore the context, we are irrelevant.

If you start with the assumption of good intention, think what might be true for someone to do or say what they did. You start to appreciate the other person's style (even if it is not your own) once you understand their perspective.

Communication Gone Awry

Dan, a developer, sent an email to Trixie, a tester, that said, "This feature is ready for you."

Stu, the Scrum Master, sent another email an hour later, "Trixie, how's it going?"

> Trixie sent off an email to both Dan and Stu, "I'm going as fast as I can. I'd be faster if I'd received the feature earlier."

We've been in the middle of email chains like that. In this case, Trixie felt under pressure for everything she needed to do. Dan wanted to make sure Trixie knew he was done. Stu wanted to know if Trixie understood everything. Everyone had good intentions for the work.

However, because email is asynchronous (not real-time back and forth) and low bandwidth, Trixie read different intentions into Dan's and Stu's comments. When people communicate asynchronously, it is easy to assume the other person meant something else when you are only considering your own context.

 If an interaction online puzzles you, pick up the phone or get on a video call to make sure you understand the other person's context.

When team members assume good intention, they give people the benefit of the doubt. Team members don't placate each other, saying what they think others want to hear. Team members don't blame others, assuming bad intention. And, team members remember that each other is human, not an automaton.

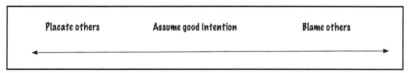

Figure 2.6: Assume Good Intention Principle

If you're curious about how to introduce this idea to your team, consider showing the team the congruence image and explaining the stances (blame, placate, super reasonable, irrelevant, congruent). Explain a recent time when you were not congruent. Now ask, "Have you seen yourself act in any of these ways?" Finish the conversation by having a discussion of what's acceptable in your team and what's not. You might want to add this to the activity in *Identify the Team's Values* on page 154.

When managers also assume good intention, and practice empathy, they will realize the team may have impediments that they can't see, and they will serve the teams so they can be more effective.

2.6 Create a Project Rhythm

No matter what style of agile practices you are following, rhythm at the team, program, and organizational level is very important. They do not have to be the same rhythm, but they should occasionally synchronize to inform each other.

Consider your team's needs for a cadence of planning, demos, and retrospectives.

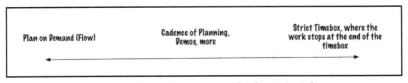

Figure 2.7: Create a Project Rhythm Principle

Some teams can use a flow approach because they are able to create small stories and release them every few hours or every day. Everybody sees the current state of the product. Experienced agile teams with sufficient hours of overlap are able to manage the planning on demand.

Some teams can use a strict timebox, where the timebox defines the planning and the releasing. The more hours of overlap the team has, the more the team can use a timebox.

Timeboxes are one way of creating a project rhythm.

Other teams use a cadence of planning, demos, and releasing. When a team uses a defined cadence of planning, demos, delivery, and retrospectives, they may be able to manage fewer hours of overlap.

The difference between a timebox and a cadence is that the team does not use the timebox to define the incoming batches of planning. The team might plan and replan at various times during the cadence. However the cadence helps other people in the organization know when to expect a public demo. For more information, see *Create Your Successful Agile Project* ROT17, or *Thinking About Cadence vs Iterations*.[1]

If your team is part of a larger effort, such as a program, your team might need to plan and retrospect more often than the program as a whole. We have seen this especially true for distributed teams. The people need a chance to reconnect more often to make sure they are working on the relevant work, finishing it, and reflecting on their efforts.

Cadence Provides Reconnection

The cadence provides the team members a chance to reconnect with each other. Team members with less than six hours of overlap may feel very isolated in a distributed team. By setting a cadence of collaboration times, these team members feel they have opportunities to reconnect with their team as humans and not just team members. Therefore, rhythm is not only good for the work but for the people on a distributed team.

See *Adapt Practices for Distributed Agile Teams* on page 173 for a more thorough discussion of planning, managing, and receiving feedback about the work.

2.7 Create Resilience with a Holistic Culture

One of the nice things about being able to work wherever you are is that people create a holistic approach to work and life. In fact, one of the lean principles is "See the whole."

---

[1] https://www.jrothman.com/mpd/agile/2017/04/thinking-about-cadence-vs-iterations/

In our experience, that holistic approach allows or encourages people to deliver their work, learn more about work, and balance their personal needs, such as family. We refer to this as work-life blending. We actually find that taking a holistic approach to work and life helps team members optimize how they work together.

I Don't See My Team Members' Families

A developer said to us, "I'm a developer. I don't want to see my teammates' families. I don't want them to see mine. I don't want the video on. I want to do my work and enjoy the fact I don't have a commute."

We understand this thinking. It works well for individuals. We have found better results in the teams we coach and facilitate when team members do take a holistic approach to their teammates.

If your organization's culture reinforces resource utilization, getting to know someone's person context is not that valuable. They either deliver or they do not.

However, if your culture reinforces flow efficiency thinking, the team members tend to appreciate each person's context and adjust their work to maintain flow.

Because of work-life blending, we have found that the teams who work best together take the time to communicate—a little—about their personal lives. When people only communicate about the work, they don't create rich relationships. When people communicate about the work, personal events, and how the other person grows, we see richer communications and better teamwork.

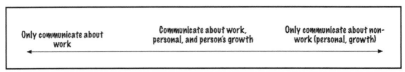

Figure 2.8: Holistic Culture Principle

As an example, one of the people on your team lives in the mountains because she likes to ski. During the winter, she takes time in the middle of the day to ski a few runs on her local mountain. People can take time during the day to see their child's school play—and the rest of the team understands.

People can create or choose a personal work environment that works for them and the team.

Any team member is able to say, "I need to leave for a while. Expect me to return at this time." The team is fine with that. The person and the team can choose what to share about that condition.

Consider these questions for understanding each other's personal context in a distributed team when your team might *Build Respect with Working Agreements* on page 147:

- How can the team respond and manage the work, regardless of someone's temporary inability to deliver their work? That's where distributed teams find value in creating a culture of team-based resilience.

- How much context is a person willing to share about their personal situation so the team members understand they need to provide flexibility for that person? People need a little time to manage their lives. Team members don't need to share every detail. They do need to share enough context so the rest of the team can empathize and adapt to the current situation.

- How temporary is this situation? If a team member's refrigerator dies, that's inconvenient and temporary. The person needs some time to deal with it. However, sometimes team members have long-term problems, such as hurricanes or fires destroying homes, and people need to find short-term relief while they look for long-term solutions.

- What happens when a child or another family member enters the view of your camera during your online meeting? Do you treat it as an interruption? Or, is it an opportunity to briefly introduce your family to your co-workers? (Hint: Mark likes to say that on a distributed team, every day should be "bring your kid to work" day).

- What happens when the situation may be beyond their control? Distributed teams might even find their distribution advantageous. Both of us have been in situations where the local weather for the entire collocated team prevented people from working—such as snow storms or hurricanes. However, distributed teams in similar organizations were able to continue.

Distributed teams need resilience to continue delivering, often in different ways from collocated teams.

2.8 Default to Collaborative Work

It might sound funny to have collaboration as a principle for a distributed team, but the agile principles require collaboration. When you use the principle of collaboration, you reduce the bus factor, increase everyone's knowledge, and create more value.

Bus Factor

What would happen if one of your team members got hit by a bus? The bus factor is the risk of not sharing information among all the team members. When a team has a bus factor of one, only one person has some vital piece of information.

Too often, distributed teams encounter the "Silo of One" (isolated-by-design) problem. People work alone, not collaborating. Developers don't collaborate with other developers or testers, the database guru or

Figure 2.9: Default to Collaborative Work Principle

the user interface expert is off in outer space, and the product owner only pops in to demand that the team take more in the backlog. When the team is under pressure to deliver, a common response is to be "heads down" and "do your work" solo.

But why would you *not* collaborate? If team members are heads down, do they see the big picture of the work? Do they check in with each other to make sure no one is stuck? Do they have an opportunity to bounce alternative ideas off each other to make the best choice as determined by the team and not an individual team member? Using the "wisdom of the team" to finish work can be far more powerful than working in "silos of one."

The more the people collaborate, the more effective they are as a distributed agile team. Collaboration allows them to understand the problem space and the possible options available to them. Collaboration allows them to learn together instead of keeping knowledge in individual silos. Collaboration allows them to explore risks together and choose the options that deliver value quickly and with high quality.

In addition, remember that agile approaches don't reward team members for *individual* work. Instead, agile approaches reward the team for finishing value through delivery.

We have seen team members mob, pair, and have successful meetings to workshop stories, create the backlog, and perform retrospectives. We'll say more about this in *Collaborate on the Work to Move it to Done* on page 185.

When people default to working together, people start to build similar personal connections that collocated teams have.

Provide Multiple Contact Information for Each Person

We have been part of geographically distributed teams where everyone had alternate phone numbers and emergency phone numbers. When Jack, one team member had a power outage, that team member could text someone else on the team. The team could still do their work, even though they were missing one person.

Without this information, the team would have been confused. Where was Jack? What was the problem? In the absence of information, people sometimes make up stories that they think fit the context. Fortunately, most of the time, these stories are wrong.

Collaboration also leads to resilience. When team members default to collaboration, they see each others' patterns. Then, if one of the team members notices that the pattern breaks, the team members can become curious and decide what to do next.

When teams are resilient, they can experiment with the product and the process. They can create and consider multiple options. The team members can help each other, in new and different ways. But the team cannot change product and process without considering business outcomes.

When managers are resilient, they focus on clarifying the business outcomes and serving the people to achieve those outcomes. The agile manager becomes a servant leader to the distributed teams and the business.

2.9 Now Try This

Now that you've seen the principles, assess your context:

1. Assess your team on each of the principles. Is your team currently using these principles to create a system of work that satisfies their needs as an agile team? If not, we'll address each of the principles in the following chapters.
2. Consider asking your team members to self-assess on each of these principles. Use some team time to discuss both the the

principles the team thinks work fine for them and the principles the team has concerns about.

3. Consider discussing the principles with your management to help them understand how to support distributed agile teams.

You can see how these principles help agile teams work as a team. Successful distributed agile teams have sufficient hours of overlap, which is why it's our first principle. Now, let's discuss a too-often occurrence: distributed teams without sufficient hours of overlap.

CHAPTER 3

Avoid Chaos with Insufficient Hours of Overlap

The first principle for successful geographically distributed agile teams is to establish acceptable hours of overlap. If your team has enough hours of overlap, it's possible to use smart people in locations all over the world. The key is enough communication time and sufficient communication tools.

Too many teams who want to use agile approaches are dispersed across the world, in a way that they cannot easily find hours of overlap.

An agile team requires a minimum of four hours of overlap a day for sufficient collaboration. Teams with insufficient hours of overlap—fewer than four hours of overlap—can avoid chaos if they consider and apply some of the approaches in this chapter.

If having fewer than four hours of overlap is your team's reality, you have choices. Even more important, your managers have choices.

First, let's examine how we got here. Then, let's look at what options you have to survive the chaos of insufficient hours of overlap.

3.1 Defining Working Across the Globe

Back in 1997, as described in *Global Software Teams: Collaborating across Borders and Time Zones*, ERR99, IBM had this great idea: they could have people on a team work in all corners of the world. One person would complete his or her work for the day and hand it off to another person several time zones away.

It sounded like a great idea. In fact, it was such a great *idea* that many organizations created geographically dispersed teams to implement that idea.

Most of those efforts failed. We have heard of one or two shorter-term efforts succeeding. Neither of us is aware of any sustained successes at this time. Also, we are not endorsing this approach. However, if you are in this situation, we hope to provide guidance that will make your team more effective.

There are two defined approaches to working 24/7: *Around the Clock* (below) and *Follow the Sun* on page 44.

3.1.1 *Around the Clock*

Organizations often use Around the Clock to provide *coverage* 24/7. Call centers might be the most common example of this. We've heard of radiologists and other asynchronous professions that use this approach.

Note that Around the Clock is not for product development. It is not for team-based knowledge work.

We use "groups" and not "teams" to emphasize that the members of this group mostly work independently. Members of workgroups do not depend on each other to deliver.

Around the Clock is for people and workgroups where people provide a knowledge-based individual service. Independent support groups, groups with enough internal capability, are a good example of this type of work.

Workgroups can use agile approaches, although not in the same way as a product development team. For more information, see *Create Your Successful Agile Project*, ROT17.

If you need to provide a service, Around the Clock might work for you. Especially if the work doesn't require coordination among the workgroup, the service providers.

If you need coordination for the outcome, consider the handoffs you might need to be effective with Around the Clock. The handoffs might make your Around the Clock appear to be more like Follow the Sun.

Around the Clock for Support

A software development manager in a large, multinational organization said this:

"We have six different support groups, who can cover most of our customers with enough overlap in terms of customer time zones. Each support group has Tier 1 and Tier 2 people. A couple of the groups has a Tier 3 person, but most of the support groups escalate to us, as a product development team.

"Each group works independently and each person in each group works independently. If it's time for a person to go home, that person can transfer the ticket to the next-west time zone support group. We have a policy of telling the customer and escalating that ticket, so the customer knows what's happening with their problem *and* the next support manager and team know, too. We have found that the transparency between ticket transfer helps.

"Sometimes, our support people need a conversation with the handoff. We set up the call centers so every center has at least one hour of overlap with the next center. We're not perfect, but we more often than not can transfer the ticket and everyone's happy.

"If we get an escalation, we put that in our urgent queue and assign it to an entire product development or feature team, so we can resolve the customer issue as fast as possible. We might not stay in sync with the originating call center, but no one has to time shift to provide excellent service."

If you work with a product development team members located all over the world, consider Follow the Sun.

3.1.2 *Follow the Sun*

Follow the Sun is exactly as it sounds: People start work in the East and the work follows the sun around the globe.

Back in the 90's when this started, internet access speeds varied all over the world. Not everyone had even access to the source code or tests or the tools to manage the source code or tests. Teams limped along, trying to make management's vision work.

Now that internet access is more available and speedy all over the world, managers think that agile teams can use Follow the Sun.

We know of several teams now, who do use a form of agile follow the sun. Here is what they do:

- They locate the Product Owner/Customer close to the eastern-most team. We mean eastern-most in terms of who is closest to the international date line—when it's Tuesday in the West and Wednesday in the East.
- They specify handoffs for each team member. As a team member completes their work, they know who to hand off that work to. Sometimes, that person is in a close time zone. More often, we see people hand off to someone more than six time zones away.
- They keep the teams relatively small, not more than a total of eight people.
- The teams who are most successful have very small stories, not more than two-day stories.
- The teams map their value stream, to identify bottlenecks as they proceed.
- The teams monitor their WIP to make sure the team isn't stuck somewhere.

While we have not seen many teams truly implement Follow the Sun, we have seen teams who specify handoffs and maintain an asynchronous backchannel have more success than other teams.

Every "agile" team we know who has tried Follow the Sun has encountered these problems:

- The team doesn't collaborate on the overall design. Instead, one person creates the design and hands that design off to the rest of the team.
- The team has trouble managing the size of their work. Too often, the team has a week-long story by the time the story is done circling the globe. (Remember: Follow the Sun was originally developed for speed.)
- The team members, because they are totally dispersed teams, often have local managers who interrupt the team members with other, non-team work.
- The team members don't plan together.
- The team members don't demo together.
- The team members don't reflect/perform retrospectives together.

This lack of collaboration creates team members who are not generalizing specialists (i.e., T-shaped people). That's because the team members don't have regular opportunities to learn from each other and share product and functional knowledge.

Unless these teams meet in person with each other at some point during the year, the team does not have the kinds of relationships with each other as in *Create Resilience with a Holistic Culture* on page 33.

We have seen supposedly agile Follow the Sun teams evolve into a terrific staged delivery lifecycle team without collaboration. (See *Manage It!*, ROT07, for more information about life cycles.)

When we speak with people on supposedly agile Follow the Sun teams, we hear stories of pain and suffering. Too many developers and testers feel as if their "team" beats them up for not delivering. Too many first-level and mid-level managers realize these "teams" aren't teams and aren't working the way the senior managers want. And, too many senior managers don't understand why the organization is still spending so much money on product development costs. Other authors have reported similar challenges with Follow the Sun. See *I'm Working While They're Sleeping*, ERR11.

Follow the Sun assumes a factory- or manufacturing-approach to product development. Software is a learning and collaborative activity, not factory work. Agile approaches use double-loop learning, where, as the team creates the product, the entire team learns and refines not just their process for their work, but the actual product itself.

Follow the Sun is anti-agile. It is anti-learning.

If you are in a team like this, you can manage and maybe survive your team's time offsets using agile principles.

3.2 Map the Value Stream to Visualize Cycle Time

Value stream mapping is a way to visualize all the work states: which people work when; for how long; and where the wait states are for your team. (*Lean Software Development*, POP03, first showed this simplified approach to a value stream map.)

Here's how to create a map:

1. Start with the very first state for the work item. That work starts above the line because the person doing the work adds value.

2. Now, count the time the item stays in that state with the same person. Keep the line in the value-add state, above the neutral line.

3. Once the first person transitions the work to the next person, decide if there is a delay or if that next person can work on the item right away. If there is a delay in adding value, that delay is shown below the neutral line. Once the next person adds value, we show the work above the neutral line.

4. Continue looping between numbers two and three until the team marks the work as done.

Here's a value stream map for an entire team *inside* four hours of overlap. The team takes advantage of their time offsets to finish the work.

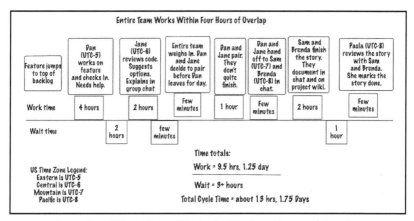

Figure 3.1: Whole Team Works Inside Four Hours of Overlap

The work happened to start in the East, with Dan, the first developer in this story. Dan worked alone on the story for four hours. Finally, he checked it in and asked for help.

Jane, the next developer with expertise in this area, happened to be in the Central time zone, one hour behind Dan. She suggested several options in the group chat, so other people could also add to the discussion. By now, everyone was at work, so the entire team could participate.

Dan and Jane paired for an hour. At that point, Dan left work for the day. Jane was ready to leave, also. They had enough time to hand off work via chat to Sam, another developer in Mountain time, and Brenda, a tester in Pacific time to pair to complete the story. Sam and Brenda worked two more hours on the story.

Paula, the Product Owner, works in the same building as Brenda, but on a different floor. It took her an hour to get time to review the story.

The story took one "team" day because of the time zones. However, the story took 13 or so team-hours, which is why we measured it at about 1.75 days. The team had a wait time of 3+ hours, for a total cycle time of 1.75 days.

The 1.75 is too precise for many teams. We recommend you use a precision of only half-days when counting cycle time. Cycle time precision often provides a false sense of accuracy. Because too often, the delays vary from story to story or day to day. Having a more accurate cycle time—here, two days—provides everyone involved a more realistic estimate. See[1] or *Manage It! Your Guide to Modern, Pragmatic Project Management,* ROT07, for more information.

We've seen many distributed and dispersed teams fall into a trap of too much WIP when they have long delays. We've especially seen Product Owners fall into that trap.

For this example, let's consider the idea that the PO will generate stories that the team members will review for the next bit of work. Your first state for a story might be to have an architecture spike, or to generate tests (system or unit tests), or to start development as your first state.

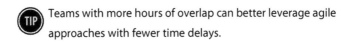 Teams with more hours of overlap can better leverage agile approaches with fewer time delays.

The team with Dan, Jane, Sam, and Brenda were able to complete their story before the west-most person left for the day. Don and his team needed more than five days just to get the stories ready for the team.

Use value stream maps to see delays. You might also see a team's WIP or the possibility of WIP, and bottlenecks. Once you see these problems, you can work with the team or management on ways to reduce your team's cycle time.

---

[1] https://www.mathsisfun.com/accuracy-precision.html

3.3 See the Effects of Hours of Overlap on Cycle Time

Now that you've seen how to map a value stream, consider this map. This team has all the product developers and Product Owner in Boston. The one lone tester is in India.

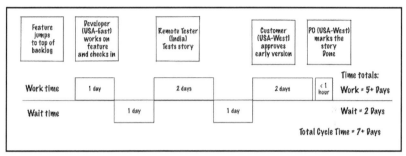

Figure 3.2: Value Stream Map

Notice that the developer spends one day on the story, Story2. This team's management thought that since the developer was in the U.S., the tester could start to work the very next day on the story. However, the tester was still working on a different story, Story1, finishing that one. The tester was new to the product domain.

The second day, the tester was able to take Story2. The Product Owner and developer had not written down enough about the story for the tester to know exactly where the story fit into this feature set and the product.

The developer had not shared his unit tests with the tester, so the tester couldn't review those tests to understand more about the story. The tester and developer had no normal work-day hours of overlap. The tester needed two days to create automated and exploratory tests to fully test Story2 and know that the testing was complete.

The Product Owner was busy with other work, so it took a day for the PO to review the story and show it to a customer.

The customer was busy, and provided comments over a period of two days to the PO. The PO and customer agreed this story was done. The PO added more stories to this and another feature set.

The work time was more than five days. The wait time was two days. The total cycle time was more than seven days. The customer injected more wait time than the team expected.

This team had no feedback loops for this story. There was no feedback from the tester to the developer or PO. Also, there was no feedback from the customer to the team.

Yet, the nominally three-day story took five days. Not because anyone on the team was "at fault," but because of the lack of synchronicity. This is the problem of insufficient hours of overlap. The value stream map shows the wait states and can help a team see potential alternatives.

The map helped this team recognize its risks and manage them:

- The team needed more story definition or documentation.
- The PO was serving two teams because he was waiting for feedback from this team. He decided to spend his time writing more story documentation and slicing smaller stories.
- The team decided to time shift once a week to refine stories and replan.
- The tester never saw a demo. The PO asked the customer and the team to schedule a demo once a month a few hours early. That way, everyone on the team could see the PO demo to the customer and the customer's reaction.
- The team had trouble finding time for retrospectives. They decided to first create asynchronous retrospectives and then to time shift once a month for a synchronous retrospective.

These team members only time shifted a maximum of twice a month. They were not a well-formed, high performance team. However, they were able to use the agile principles and create a better team and a better product as they proceeded.

This team was able to time shift to start at 8:00 A.M. in Boston, 5:30 P.M. in Hyderabad. That time shift was not horrible, although inconvenient. The PO ordered gallons of coffee to help people be awake.

The real issue was the commute time of the Boston team members. They normally arrived at work closer to 9:30 or 10:00 A.M. each day, to manage their commute time. The time-shifted 8:00 A.M. start meant that several people had to leave their homes closer to 6:30 A.M., much earlier and with much more disruption to their personal lives. The team members didn't like that.

One alternative for this team would be to treat the entire team as if they were a fully dispersed team, and create a virtual workspace as in *Create Your Collaborative Team Workspace* on page 111. When the team assumes each person is remote, no one has to physically arrive at work by 8:00 A.M. That virtual workspace provides more hours of overlap.

At the time, management objected to the perceived few thousand dollars of total expense to set up the virtual workspace. The value stream map showed management that the team wasted three days on every story. As a seven-person team, those three days cost approximately $10,000/week. Once the team showed management that they wasted about $10,000 each week, the few thousand in expenses to outfit each person on the team with sufficient bandwidth and licenses for the tools they needed seemed small in comparison.

You have options for managing work across insufficient hours of overlap.

3.4 Manage and Survive Your Team's Large Time Offsets

What can you do, if your team is a global team covering many time zones and insufficient hours of overlap? You have options:

- Develop working agreements for your team.
- Train each person to be able to use and access all the tools.
- Understand what it will take for your team to start with the east-most person for the work. Map your current value stream and decide if you need to do anything different.
- Use time shifting by choice and sparingly.

While we do offer some suggestions here, realize that all of the information is for you to make the best of a bad situation.

We'll start with working agreements.

3.4.1 Develop Working Agreements for your Widely Dispersed Teams

When people have so few hours of overlap, we often use these kinds of working agreements:

- Document the problem and the solution in the code and tests.
- Create pictures with words so people who need to *see* a design or an architecture can see it. The image and the words help other people on the team understand what is in each person's head.
- Create a problem escalation process for the team that does not require management intervention. Problems that have to go up and over via a management hierarchy tend to introduce delays.
- See *When Each Person Works* on page 153 to see *when* everyone is working. Then, if a team member needs to try synchronous communication, each person understands when that communication might be possible.
- Define what each person will hand off to the next.

Notice that the team focuses on documentation, not verbal collaboration.

Dispersed teams with no hours of overlap are rarely successful using agile approaches.

There are three key questions for teams with no hours of overlap:

- Are these teams survivable short-term? Maybe.
- Are these teams sustainable? Not often, not as a team. Rarely as an agile team.
- Can teams like this deliver sufficient value that the cost of trying to work in a team like this is worth it to the team members and the organization? Probably not.

We have seen too many teams feel stuck in situations like this, where the product they produced never had any customers. That's because the Product Owner (or Customer) was not able to interact sufficiently with the rest of the team members. And, the team members were not able to collaborate sufficiently to build on their learning to create a better product.

3.4.2 Train Everyone to Access and Use All the Tools

Does everyone on your team have access to each tool? We have seen two forms of access problems: insufficient licenses and insufficient training on the tools.

The problem with sharing licenses is that it only takes one forgetful day to destroy trust and prevent other people from progressing on their work.

When each person on a team does not have the same tool access, everyone loses. The team member without access loses time, waiting for someone else to kick off a build or test. The team loses the increased knowledge from that team member. Also, the team members who can start builds or tests often lose time when other team members request those builds or tests. The team members with access experience multitasking and the cost of switching between other tasks.

 Teams lose days of work waiting for others when people don't have access to licenses and capability in using the tools.

Make sure everyone on the team has equal access and knowledge for all team tools.

Who Has Tool Access?

One dispersed team had one build person in Boston. The rest of the team was dispersed throughout the rest of the world. Only the Boston person could start a build.

That team had a tester in San Francisco. Only the tester could start the tests after the build.

That "team" experienced at least one day of delay every time someone in China and India changed the code. When the team finally mapped their value stream, they realized their cycle time was a minimum of four days, with three days of wait time. In addition, their WIP was enormous.

That data allowed them to work with the relevant managers and buy sufficient licenses and training so everyone could start the build and the build tests.

Managers too often confuse wage cost with project cost. See *Think in Flow Efficiency* on page 10 for more information and *How Leaders Can Show Their Agile Mindset* on page 227 for more information about costs in distributed agile teams. When the team maps its value stream and measures its average cycle time, the team can make the financial case for equal access and training for tools.

3.4.3 *Launch the Work from the East*

When teams have many time zones and little to no hours of overlap, it makes sense for the *team* to start work from the East. The starting point begins with the person closest to the international dateline.

Don, a Product Owner, works out of San Francisco. His team consists of four developers in New York and three testers in Singapore.

Don works with the developers to workshop stories. Once the developers are satisfied with the stories, Don time-shifts at least once a week to work with the testers to review and refine acceptance tests for those stories. Because Singapore is 15 hours ahead of San Francisco, Don works at 5:00 A.M. on the days he works with the testers in real time. The testers work at 8:00 P.M. on those days. Their value stream for creating the stories—not completing them, creating them—looked like this:

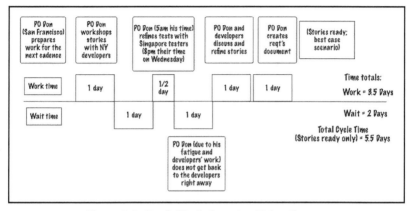

Figure 3.3: Don's Workshopping Value Stream

Don often iterated a couple of times with each group to finish refining the stories.

Don created written requirements for the team to work on for the next couple of weeks. This team used a planning cadence for their project rhythm, not a strict timebox.

The testers started the story first. They wrote system-level automated acceptance tests. They also wrote automated tests at the feature level.

The developers received the tests. Some developers wrote unit tests first (as in TDD or BDD), and some did not. They often had questions for the testers, even though everyone had access to the requirements document.

Don then measured the cycle time for his team. Once the testers receive the story, each story takes an average of two days to complete—two circles of the globe. Add in his time iterating on preparing the story, and the real cycle time was closer to four days per story.

Don used this data to request his management consider creating cross-functional teams in each location, instead of a dispersed team. Don explained about the personal cost to him and the testers of his early work and their late work. Don now has a copilot Product Owner and they share the time-shifting pain.

This is a map of how the team worked to finish their stories:

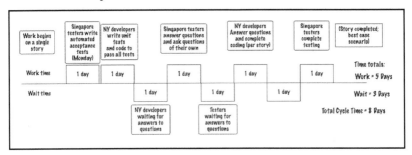

Figure 3.4: How Don's Team Worked to Complete Stories

Everyone here felt the pressure of the time zones and no hours of overlap. When this team launched the work from New York, the cycle time was even longer. That's because the team had not fully documented the requirements.

One other problem with waiting for answers to questions is that people start something else while they wait. That means that every single person on the project multitasks—even if the work is for the same project. Every person incurs some switching cost when they return to the original work. (See this post to visualize this cost of delay.[1])

The only people who did not feel the immediate pain were the managers.

 If your managers are not aware of the pains of distributed work, do what some police forces do: offer a ride-along program for a day or two. Help your managers feel the pain.

If your team has these time zone problems, consider gathering data:

- Map the value stream for your team for the most recent week or two of work. The value stream map will show the wait times.

---

[1] https://www.jrothman.com/mpd/portfolio-management/2014/02/cost-of-delay-multitasking-part-2/

- Measure the cycle time. We recommend you consider measuring the shortest cycle time, longest, and median cycle times. That helps the managers see the best case, the worst case, and the more typical case. An average cycle time might not be the same as the median.
- Measure the team's WIP. Too few managers understand that WIP saps a team's ability to feel as if they finish anything. Too much WIP and people stop finishing, even if they are just an hour away from finishing, (as in *The Progress Principle*, AMA11). If you don't look at the queues where work waits for the next stage, you will miss valuable insights.
- Measure the cost to complete a feature, over the last several features or the last couple of weeks. Calculate the cycle time in days for those features. Calculate the daily run rate for the team. Now, for each of those features, multiply the run rate by the cycle time, and you have the cost to complete those features.

Make the data visible to the entire team and managers to gain insights on how to change the work organization or the team locations or the flow of work.

When teams use agile approaches, the question is not about which people have the work. The question is how does the team move the work to done as effectively as possible.

That's why we recommend you start work in the East. You might need to see your team's work states and wait states. That's what a value stream map can show you.

3.4.4 Time Shifts Can Help With No Hours of Overlap

When people shift their work hours to create hours of overlap, they time-shift. While we don't think it's reasonable to ask people to permanently shift their time, sometimes people want to shift. And, if they are willing to do so, time-shifting can create some hours of overlap.

When people time-shift, they might reduce delays for the team. In Don's story *Launch the Team's Work from the East* on page 54, Don time-shifted to create opportunities to collaborate with the Singapore team. However, Don was unable to sustain more than three hours of time-shifting.

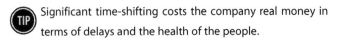 Significant time-shifting costs the company real money in terms of delays and the health of the people.

The personal costs of time-shifting can be quite high, in personal interruption and health issues. We've met people who regularly wake up at 3:00 A.M. to have a conversation with someone across the globe. Their company expects them to put in a full day of work (in their normal time zone), after their interrupted sleep.

These employees make these personal sacrifices because they feel invested in the work and may push themselves to the point of exhaustion. If they do not feel the company is "protecting them" from this unsustainable pace, even if self-imposed, they may no longer feel it is worth investing additional time into their work. They may even begin to devalue their work and the company.

For these reasons, we recommend that companies do not *depend* on people time-shifting to create successful projects.

People can time-shift for a short time to help a project, but people cannot sustain significant, mandated time-shifting. Too often, these people time-split—they start very early in the day, take a break, and return to work later, often for a full eight-hour or longer day.

Some team members may even time-shift by choice because the hours of overlap needed by the team happen to align with their working preferences—early or later for their time zone.

Since agile approaches are built on sustainable pace, time-shifting is a short-term solution.

When someone burns out because of time-shifting, the company loses someone with extensive knowledge and has to rebuild that knowledge and capability. Realistically, the company may only partially reclaim this knowledge and capability and they will suffer business consequences.

Instead of long-term time-shifting, we recommend handoffs. The best handoffs require some time overlap—at least 15 minutes.

3.5 Consider Handoffs

Handoffs are exactly what they sound like: one person hands off work to another person. Handoffs require trust and respect between the people.

We've seen handoffs in a hospital setting. The nursing shifts include handoff time so the nurses can discuss each patient. Your team can do the same.

We recommend you have team members who have at least one hour of overlap for their chosen working hours. Instead of pairing, the team members hand off work to each other, from east to west.

When a team member is at the end of his or her day, the team member has one more action: a real-time and as natural a communication as possible with his or her colleague in the next-west time zone. The easterly person describes the work and their progress. The western person takes that work, asks any questions, and then continues to work on that item until either the work is done, or that person hands it off to the next person.

We don't recommend email for handoffs. We recommend a conversation. If your team members have no hours of overlap, try experimenting with the form of the handoff message in asynchronous communication.

This works best with very small stories. That's because smaller stories tend to have lower complexity. And, the people involved can see where the original person started and how to finish.

We've seen handoffs fail when the western person rewrites the eastern person's work. Do that enough, and you destroy respect and trust.

3.6 Insufficient Hours of Overlap Traps

We've addressed many of the problems teams have with insufficient hours of overlap. Here are common traps you might address.

- Teams create a semi-permanent time-shifter.
- Attempting to use highly collaborative face-to-face approaches instead of adapting practices to the team's reality.
- Not using the value stream to see where it might make the most sense for work to start.

Let's examine each of these in turn.

3.6.1 Trap: Teams Create a Semi-Permanent Time-Shifter

Too often, one person shoulders the responsibility for enabling the entire team's work, even with insufficient hours of overlap. That person creates "team debt" because they take the responsibility instead of distributing the responsibility for team communications among the team.

That person used to be called a project manager. Now, that person is often called a Scrum Master. We hear these people call themselves various other names: Point-Person, Whipping Boy, and Fall Guy. (These names happen to be masculine-sounding. We know of plenty of women in these positions who use these monikers to refer to themselves.)

While time shifts can help the team, very few people can maintain personal permanent time-shifting without a high personal cost.

When the team relies on one person to shoulder all the hours of overlap responsibilities, the team incurs significant team debt. We've seen these forms of team debt:

- The time-shifter forgets, drops, or otherwise doesn't complete in time for the team to take advantage of all the team members.

- Lost opportunities for team communications. With the time-shifter "coordinating" across the team, individual team members tend to communicate less.

Instead of one person permanently time-shifting, consider these options:

- Add a copilot, ambassador, or buddy in enough locations that these people can share the pain of insufficient hours of overlap.
- Reconfigure the team so they can work as a team, with sufficient hours of overlap.
- Instead of starting with agile teams, consider starting with the managers, as in *Start with a Distributed Agile Management Culture* on page 249.

3.6.2 Trap: Assume a Face-to-Face Approach Will Work

Teams with insufficient hours of overlap encounter plenty of trouble when they try to apply a "standard" agile approach "by the book." In many cases, the "book" assumes a specific work environment where all team members are collocated and able to easily exchange tacit knowledge.

Too few organizations provide a work environment where the distributed team can easily exchange tacit knowledge. The team needs to construct a work environment that helps them work as a team, share and capture tacit knowledge, and deliver their product.

Many classic agile practices fail when a distributed team attempts to apply them *as is* without first constructing their team environment.

Instead of wholesale agile practice adoption, successful distributed teams first review the principles behind the practices. When the team understands why the principle works for a collocated team, then the team can adapt those principles for their specific circumstances. We'll address this more in *Adapt Practices for Distributed Agile Teams* on page 173.

This is difficult and intense work. Distributed agile teams, especially teams with no hours of overlap, need time to create their agile practices.

3.6.3 Trap: Not Using the Value Stream to See Where to Start the Work

Once we have some idea of the requirements, with any luck in story form, we often think we need to start with the developers. Especially if you have a factory mentality about software or other knowledge work, we think we need *something* before we test it.

However, if you see your value stream, to see where you have delays in your process, you might need to start with the testers.

We've seen too many "teams" where the Product Owners and developers are much farther west than the testers. The testing delays grow over time. The team's cycle time increases.

We've even seen teams where the developers have one "iteration" followed by a "tester iteration." This is not an agile approach of any kind. It's smaller waterfalls.

What would have to be true for your project to start with the east-most person? How can the rest of the team change to more effectively work as a team? See *Launch the Team's Work from the East* on page 54.

3.7 Now Try This

1. Map your team's value stream. Understand how many times each story—or worse, each task—circles the globe. Until you see where the various team members add value and delays, you can't know what options you have for change.

2. Consider handoffs. The smaller your team can slice each chunk of work, the more likely each team member can complete valuable work in a day of their time. Then, each team member can hand off completed work to the "next" member of the team.

3. Develop working agreements for your team. We recommend you bring the team together as in *Non-Collocated Teams Deserve Face-to-Face Time* on page 12. If you can't bring your team together, consider how the team can understand how each member prefers to work and how those preferences work for the team overall.

4. Cross-train early and often on tools. Every team member must learn and master all the team's tools. This includes each person having their own personal license for each of the tools.

5. Launch from the east, the person closest to the international date line. You will have better results if you can launch work from the eastern-most person. You may have to have a more westerly person create the ranked backlog and stories. However, if you start work from the east on a given story, the story encounters the fewest delays.

6. Start all together for at least a week, preferably two weeks. Too often, managers and finance people are suspicious of the money a team needs to gather everyone together at the start of a project. We have found that starting, even for just two weeks at a time, pays tremendous dividends over the course of the project. In addition, when you bring people together once a quarter as in *Non-Collocated Teams Deserve Face-to-Face Time* on page 12 the team members reinforce their trust of each other.

7. Use time shifts cautiously. It's easy for people outside the team to say, "It's just one meeting." Too often, we see that one meeting becomes a standing meeting, and someone pays the personal price of attempting to time shift again and again. Instead of time shifting, consider handoffs and more frequent in-person meetings in various locations throughout the world.

You might notice we've limited our suggestions in other chapters to three or four suggestions. This chapter requires many more suggestions because it's so difficult for humans to collaborate when they cannot use synchronous, rich and natural collaboration tools.

We'll cover various types of communication in *Communicate to Collaborate* on page 87.

Too many teams faced with insufficient hours of overlap feel as if they have no control over anything. They are too often correct.

Agile teams take control of their team's environment.

This is a good "test" for how much agility your managers want your team to have. If the managers encourage the teams to create an environment that works for that team, the team has a chance of setting the stage for success.

On the other hand, if the managers want to control the team and create two-week waterfalls, the team can decide what they will and won't do to accommodate the managers' desire for the label of agility versus the actual agile culture.

Next, we'll talk about the various team types we've seen, and how they can build affiliation as an agile team.

CHAPTER 4

Identify Your Distributed Agile Team Type

When you understand your team type, you can more easily visualize your team's opportunities and challenges for collaboration. You can discuss the context for each of the team members, project, and organization. You can identify the patterns of communication and collaboration that work for these teams.

Those patterns might change how managers and executives understand how to create successful teams. Each team might need a different configuration, tools, and travel budget to be successful.

Instead of lumping all non-collocated teams together, we'll describe the four different patterns we see most often. When you can see your team's configuration, you can see if you've created a team with high affiliation and better possibilities for collaboration.

Once you know the type of team you have, you might find it easier to diagnose problems, determine the type of team you actually need, and actually make progress.

First, let's discuss a little bit about the "right" size for your team.

4.1 Consider Your Team Size

There is no "right size" for any team, never mind a distributed agile team. The question for you is: how well can your team collaborate as a team to finish the work?

 We recommend a distributed team size of between four to six people, including the Product Owner, any coaches, or

Scrum Master/agile project manager. Yes, that means six people total.

One of the problems with collaboration is that the number of communication paths—especially across distance—can prevent people from collaborating. And, the number of communication paths increases exponentially as the number of people increase on a team. With N people on a team the number of communication paths = $N*(N-1)/2$.

That means you have these communication paths for these team sizes:

Team Members	Communication Paths
4 people	(4*3)/2=6
6 people	(6*5)/2=15
8 people	(8*7)/2=24
10 people	(10*9)/2=45

Figure 4.1: Team Size and Communication Paths

Once we get past 21 communication paths, many of us don't talk with the *entire* team—we choose who to work with and who to ignore. A team of 10 people, requiring 45 communication pathways, is unmanageable for most of us as a collocated team, never mind a distributed team.

Our experience is that once distributed teams get to nine people, they subdivide on their own. They might divide into eight people plus one siloed person, or seven plus two people in their own silo, and so on. This inhibits communication even further.

Four to six people often provides a team with enough communication paths and enough complementary skills and capabilities to be a full team.

Fewer communication paths might help create more team affiliation.

4.2 Encourage Team Affiliation to Improve Collaboration

Collocated agile team members and closely-located teams in an agile organization have many ways to build affinity or affiliation among people. They might go out to lunch. They might have in-person Communities of Practice or other learning opportunities. They might socialize around culture, such as movies or cooking, or some other common interest.

Distributed agile teams don't have the same opportunities to create affinity inside the team and across teams. That's why distributed agile teams may find it difficult to work according to the eight principles.

Geographically distributed and dispersed teams have communication and location challenges that collocated teams do not. These challenges are:

- How team members affiliate as a team and why that affiliation helps the team.
- How managers can help create (or destroy) affiliation.
- How communication challenges affiliation.
- How affiliation may be supported outside of the work.
- High-performing agile teams affiliate as a team.

Affiliation is a function of the container that defines the team, differences within the team, and exchanges between team members needed for a team's success. Complex adaptive systems explain these containers, differences, and exchanges.

4.2.1 Teams Are Complex Adaptive Systems

In *Adaptive Action*, EOH13, Eoyang and Holladay describe affiliation in these ways: Containers, Differences, and Exchanges (CDE). The concept of CDE allows teams and leaders to think about how team affiliation exists now and how it can shift to be more productive.

The *Container* is how people affiliate as a team.

We hope that each team has *Differences* with respect to each individual's capabilities and skills. These differences create a cross-functional team. The differences help each person contribute their best to the team.

Too often, we see differences as not helpful. Instead, consider how differences in experience can help a team. Each team member might be able to provide different contributions. New team members can learn why the team made previous choices. Each person's past experiences might help the team question and evaluate their current choices.

This is why "agile teams" with developers in one location and testers in another location don't work. "Agile teams" of one function are not agile teams. An agile team is cross-functional, and each team member has sufficient differences so the team can deliver working product, not pieces of functionality.

Team members *Exchange* information, as well as support and respect with each other on a regular basis.

> "Systems with many and/or very tight exchanges self-organize quickly, and fewer or weaker exchanges tend to slow down the self-organizing process."[1]

This explains why teams with more hours of overlap that take advantage of this overlap by defaulting to collaborative work can self-organize faster than teams with fewer hours of overlap or who choose to do more solo work.

4.2.2 *Affiliated Teams Deliver*

We each affiliate with various teams over time. When distributed team members affiliate with the *team*, not their function or manager, they

---

[1] http://s3.amazonaws.com/hsd.herokuapp.com/contents/2/original/Eoyang_CDE_18FEB11_%282%29.pdf?1332116813

are more likely to succeed. These team members remain focused on the team's work.

 Team members can be part of multiple containers, but to deliver value, the strongest container—affiliation—should be with the product or feature team.

While a team member can be part of multiple containers, they should be on only ONE product team at a time. Other affiliations support their work on that team, such as Communities of Practice. Team members might create Communities of Practice around architecture, design, testing, performance, and more.

When team members affiliate with the team, they align to the team's purpose and process, making those daily small decisions that create value on a regular basis. When team "members" are members in name only—because they affiliate with their manager—the team struggles to communicate, collaborate, and deliver.

Team size and affiliation are not the only issues that affect how well team members work together. Each team member's location and the relative distance of each location also affects how well team members can work together.

4.3 Understand Boundaries of Collocation

Dispersed teams can never be collocated, but some "collocated" teams, separated by boundaries, are actually distributed teams, even if they are in the same building.

In *Developing Products in Half the Time*, Smith and Reinertsen, ROR98, use the Allen Curve to discuss the distance at which you can consider a team to be distributed as opposed to collocated. The closer people are physically, the more they tend to communicate.

 Just because people *can* walk to each other and talk, does not mean they *will*.

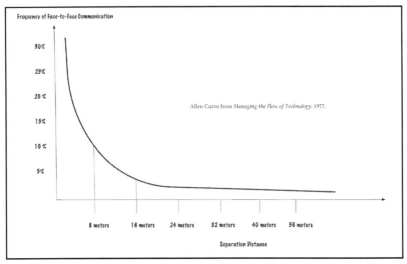

Figure 4.2: Allen Curve

The Allen Curve explains how collocated teams can be distributed teams. While 30 meters isn't very far, it's still approximately 40 steps,[2] about 30 seconds of walking to another person.

Cockburn, in *Agile Software Development*, COC06, suggests the entire team be within the length of a school bus. In the U.S., a school bus is typically 45 feet (13.75 m) long, less than half of 30 m.

That means cross-functional, collocated teams are in one location, with people fewer than 30 meters apart. We will assume that people in one physical location are willing to walk up to 30 seconds to talk with each other, up to 30 m. If your team members are not within 30 m of each other, you have a distributed team of some type. When elevators or stairs separate team members—even in the same building—the team members are not collocated. We'll call these boundaries a "geo-fence."

---

[2] http://www.kylesconverter.com/length/meters-to-steps

Identify the Cost of Asking a Question

When people are separated by 30 m or less, the time it takes to walk to ask a question is about a half minute. That person can still think about their work on their walk. However, the longer a person has to wait to ask the question, the more likely that person will become distracted by the environment or by other people around them. That creates a cost of delay due to multitasking. The more delay before people ask questions, the more questions they have, and the more the delay cost looks like a hockey stick. (See *Measure Costs of Delay in Distributed Teams* on page 228 for more information.)

In collaborative teams, we want people to ask each other questions. However, especially in a geo-fence situation, where the managers *think* the team members are close, the cost to ask a question increases.

Now it's time to identify your distributed team type.

4.4 Define Your Team Type

Each distributed or dispersed team is unique. We've seen several patterns of successful distributed teams:

- Satellite teams: the bulk of the people are in one location, while one or two people are remote.
- Cluster teams: multiple people are collocated in two or more locations, but fewer locations than the number of team members.
- Nebula teams: each team member is in a separate location from each other, dispersed.

Why not just use "distributed" and "dispersed" labels? We find that most people don't hold the distinction in mind. Also, the labels of "satellite," "cluster," and "nebula" can help you visualize the team type as well as help you recognize advantages and traps associated with each type.

As you review the team types below, keep these ideas in mind:

- A solid line between people means that they can easily communicate because they have at least six hours of overlap. Collocated teams have this advantage.
- A heavy dashed line between people means they have at least four but fewer than six hours of overlap to communicate and collaborate.
- A light dashed line indicates minimal (less than four hours) to no hours of overlap for communication.

The circle around the team represents the strength of each member's affiliation with the team.

- A solid circle around the team means all the team members affiliate with the team. The team is their "container."
- A dashed circle around the team means that some or all of the team members have a different primary affiliation, such as their functional group.

For completeness, let's start with collocated teams so we can better understand the differences between all the various team types.

4.4.1 *Collocated Teams*

Collocated teams have physical spaces in which to collaborate. They might have a team room. They might have a common physical board. They almost certainly have a whiteboard for the team to explore concepts.

The solid circle around the team represents the team affiliation. Successful collocated agile team members affiliate with their *team* over their *function*.

Collocated team members aren't always "side by side," as in a team room. However, they can easily talk with each other in real time.

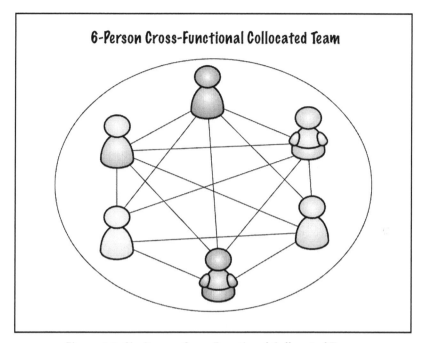

Figure 4.3: Six-Person Cross-Functional Collocated Team

Collocated team members can see each other's body language. They can overhear other team member's conversations and make immediate connections about other people's intent and the consequences for the product. Each person has a direct communication path to each team member, as indicated by the solid lines.

The team is close enough that it controls its team communications.

4.4.2 Satellite Teams

Satellite teams are cross-functional and dispersed. Most of the team is in one physical location, often collocated with each other. One or more people "orbit" around the team, in separate locations. For example, we know of a team with four team members who sat next to each other in Cambridge, Massachusetts. They also had a team member in Nashua, New Hampshire, about 40 miles to the north, and another team member in Worcester, Massachusetts, about 44 miles to the west. That team had two orbiters.

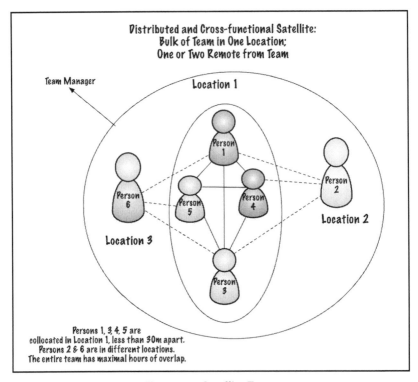

Figure 4.4: Satellite Team

In satellite teams, the remote team members (i.e., the satellites) may have trouble affiliating or even feeling connected with the team. The dashed container around the entire team on the left shows this partial affiliation. See Figure 4.5.

When teams move from partial team affiliation (a Container with part of the team), to full team affiliation (the Container as a whole for the team), the team can build respect, understand and utilize their Differences, and Exchange what they need as a team.

You might have satellite teams under several circumstances:

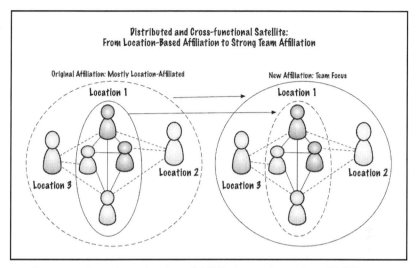

Figure 4.5: From Location Based-Affiliation to Team-Based Affiliation

1. You have a person (or persons) with domain expertise who wants to continue working in the organization and may need to work from a different location for personal reasons, and you don't want to lose them.
2. Your company bought or acquired another company and team members do not want to move.
3. Your company shut down offices in another location and integrated other people as satellite members of a different team.

You may have started with a collocated cross-functional team, and now you have a satellite team.

Not all distributed teams are satellite teams. Some distributed teams look more like clusters.

4.4.3 Cluster Teams

Cluster teams have multiple collocations. That is, the team has at least two locations where the team members sit near enough that they think they are collocated. In addition, the cluster team may have other team members who do not share any location.

Similarly to satellite teams, cluster teams may forget about the people in other locations. Cluster teams tend to form subteams.

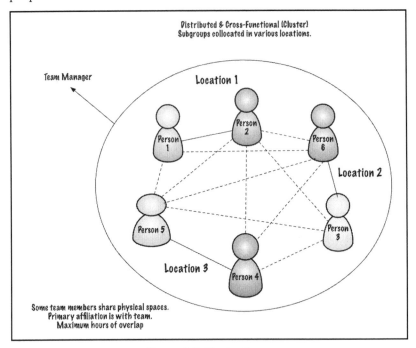

Figure 4.6: Cluster Team

In the picture, Person 1 and Person 2 sit next to each other on the third floor of their office building. Person 4 and Person 5 sit next to each other on the fourth floor. Person 3 and Person 6 work on the second floor.

Each of these pairs of people are collocated. That's why each pair has a solid line connecting them. Because each pair is in a different location, each person has a dashed line connecting the people who are not collocated.

This team affiliates as a team, not as individuals. That's why the team has a solid line around it. They all have a manager, who helps them grow as individuals and removes team obstacles they cannot remove.

Is Your Team Distributed In One "Location"?

Too often, the managers and the team members don't realize the team is a distributed team. That's because while the team members *appear* to be in one location, they are not. The team members are dispersed across one city block, or several floors of one building. However, the team members are not collocated.

If your team is separated by a geo-fence, that team is a distributed team. Even though the people might park in the same parking lot, some team members are separated by more than 30m.

Many large companies have cluster teams. Team members may all be on the same campus but in separate locations. The members of the team who are collocated will tend to talk with each other because they are near each other. The non-local team members have to use the team's virtual space.

Why Not Collocate the Entire Team?

Every so often, the team members ask, "Why can't we collocate, move everyone together instead of using near clusters?" The quick answer is organizational politics.

We have seen the Furniture Police and managers (because of their incentives and rewards) not want to move people to be one collocated team.

Too often, people are supposed to work on several teams at one time. Many organizations don't understand the significant Cost of Delay in starting and restarting a team, especially a dispersed team.

There might be a third reason: people don't want to move. They may be near colleagues and people whom they can ask questions and bounce ideas off of. Moving has several costs, some of which are personal.

Clusters can be farther apart than a single campus. We know of a company where two developers are on the U.S. East Coast sitting in the same office, two testers are in the U.S. Midwest sitting next to each other, and the Product Owner is on the U.S. West Coast. The three locations make this a cluster team because the people on the East Coast and Midwest are collocated.

You can think of cluster teams as "near-clusters" and "far-clusters." Near-cluster teams have the capability to meet in person to plan, design experiments, and retrospect. Far-cluster teams may not have the same possibilities to meet in person.

You might want to use cluster teams in these circumstances:

- You have a new project and expertise in different physical locations. You might need to ask people from different locations to collaborate on the project. Each location has expertise that people can bring to the project.
- If your company has grown by acquisition, you might have people with expertise in different locations and you want to cross-train the various people.
- You've used waterfall approaches with distributed and dispersed teams, and you want to move to an agile approach.

What if all your people are in the same building, but not next to each other? Your team is dispersed. You don't have a cluster team. You have a nebula team.

4.4.4 Nebula Teams

Contrast the distributed team, which has some collocated members, with a dispersed team, which has none.

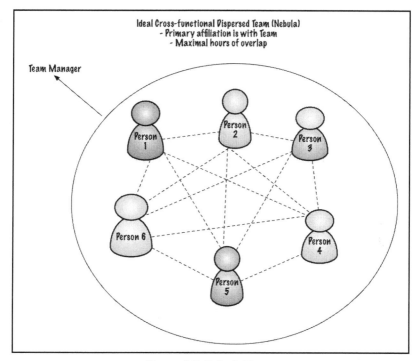

Figure 4.7: Nebula Team

Every team member is separated by at least 30 m. That is why the team members have only dashed communication lines. A solid circle encloses the team, showing that the team affiliates as a team. They also have a team manager who serves the team.

It's very easy for a distributed (cluster) team to become a dispersed (nebula) team: People work from home, or move to different floors, or move to another building. Where people were collocated as subgroups, they are now dispersed.

High-performing nebula teams are quite valuable. Team members can work anywhere.

Some nebula teams are not cross-functional teams. They are silos, which creates problems for using agile approaches.

4.4.5 Siloed Teams

Team members who sit with *and affiliate* with their functions, rather than their team members, have multiple opportunities to learn from each other every day. However, the team's workflow often suffers. The team members need to learn how to work primarily with their team members, not with their fellow function members.

In this image, the darker dashed lines show people who affiliate more with each other because they have sufficient hours of overlap for collaboration. The lighter dashed lines show insufficient hours of overlap for collaboration.

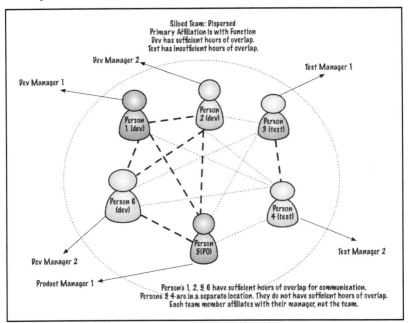

Figure 4.8: Siloed Teams

This image shows how one siloed and dispersed team might look. Person 1, a developer in San Francisco, worked closely with Person 6, a developer in San Jose. Person 2, a developer in Boston, was also able to work with a Product Owner (Person 5) in Denver. Because all the U.S.-based people had sufficient hours of overlap, they were able to communicate fairly easily.

However, Person 3, a tester in Kiev, and Person 4, a tester in Bangalore, could only easily collaborate with each other. They had insufficient hours of overlap with the developers and Product Owner.

Each person reported to a different functional manager. Each person was more affiliated with their functional manager than they were with their team.

Here's how to make the silos work:

1. Make sure the people who are supposed to collaborate across distance are relatively senior in their functional skills and self-sufficient in their domain expertise. (Functional skills refers to a developer's knowledge of development, a tester's knowledge of testing, etc. Domain expertise refers to how well the person understands the internals of the product in development. See *Hiring Geeks That Fit*, ROT12 for more details.)

2. If the people are not relatively senior nor relatively self-sufficient, they need a buddy. These people might buddy with someone locally for a while, or buddy with people on the team in some other location. We have seen the buddy system work well when people *Use a Buddy System* on page 142 for a minimum of two weeks.

3. Watch for anyone thinking in "utilization" terms. When managers start to think about utilizing people, they ask team members to multitask. That multitasking will incur delays on your project and prevent people from collaborating. See *Lead Distributed Agile Teams to Success* on page 225 for more detail.

4. Watch for hours of overlap. If your team has very few or no hours of overlap during the regular work day, the team members will not affiliate primarily with their team. They will affiliate with their function.

If you discover that either through turnover or some other change that the people on your team are not relatively self-sufficient, act as soon as possible. We recommend product measures of progress as in

Create Your Successful Agile Project, ROT17 to see how the team progresses on their work.

You might see more than one team type in your organization. If your organization is large, expect different team types.

4.5 See Your Team Type Traps

Your team may encounter traps based on its team type. Here are some common traps we've seen:

- Satellite teams have trouble remembering every member for all their work.
- Cluster teams may not work as a complete team.
- Nebula teams may affiliate by function instead of by team.
- Siloed teams may not be able to create cross-functional teams.

Your team may encounter a combination of these—or other—traps based on the team type. If your team is large, you may recognize traps from multiple team types.

4.5.1 Satellite Team Traps

It's easy for people in the main part of a satellite team to forget about the one or two people who work in a different location. Here are several instances we've seen:

- You have a contractor with expertise, but no prior relationship with the company or the team. We too often see the team not be able to integrate this one person with the rest of the team. They tend to focus on their own work that requires their expertise.
- The "remote" versus "headquarters" label. Make sure you treat the satellite people as bonafide full-class members of the team. We will say more about this in *Language Matters* on page 104.
- When a team shifts from collocated to distributed, a team may forget to shift their practices and this can damage collaboration.

We will say more about this in *Adapt Practices for Distributed Agile Teams* on page 173.

Address these challenges with: enhanced communication (see *Enhance Discussions with Dedicated Backchannels* on page 100), a shared team workspace (see *Create Your Collaborative Team Workspace* on page 111), and by *Using a Buddy* on page 142 system to help include the people in different locations from the main team.

4.5.2 *Cluster Team Traps*

Because some people are either close or collocated in a cluster, the entire team might not work together. Sometimes, they develop cliques based on location. Here are some of the examples we've seen:

- We forget to talk to team members in other clusters, other physical locations. We don't realize other people don't see or hear the social and communication cues used during in-person communication.
- People on the team forget to collaborate with each other to actually transfer the knowledge. Limited hours of overlap might challenge the collaboration abilities of the people on the team.
- One location might not initiate/respond to other location(s). We often see Conway's Law (as in *The Mythical Man-Month*, BRO95) with cluster teams.
- The team might actually be siloed. We know of too many teams who were cross-functional on paper but did not collaborate enough. They needed to use the hierarchy of their management and project management to raise issues, challenges, and problems.

See *Communicate to Collaborate* on page 87 for specific suggestions on how your team can use a variety of communication systems and tools to manage their distribution. In addition, see the ideas in *Create Your Collaborative Team Workspace* on page 111.

4.5.3 *Nebula Team Traps*

Because nebula teams are fully dispersed, the team members may experience a pull away from their teams and toward their managers. These are some examples:

- Team members may affiliate with their functions rather than with their team. These teams are siloed teams.
- The communication challenges with the nebula teams tend to be in these areas: hours of overlap, imbalance of deep focus vs. collaboration time, and personal/family life interaction. Teams can address these issues with working agreements. See *Develop Working Agreements for your Dispersed Teams* on page 52.

To manage these challenges, make sure that the *team* has sufficient hours of overlap, that the team creates its workspace, and that the team has a team focus on its work. Also consider taking time to *Build Respect with Working Agreements* on page 147.

4.5.4 *Siloed Team Traps*

Too many managers say, "You, you, and you—you're a team!" Teams learn to become teams by working together. That means the team learns to trust each other by collaborating and assisting each other to deliver completed work.

Too many siloed team members feel pulled away from the team, to their functional group.

- People tend to communicate and collaborate with the people physically closest to them. In this case, it's their functional team colleagues. That means that the team container is weak and the functional container is strong.
- Siloed teams have many communication challenges. Because these team members often are physically close to their managers, the managers might ask the team member to work on something other than the project. Sometimes this is to "fully utilize" the

person's time. Also, if the hours of overlap are small, the team member might not realize he or she doesn't understand the work and then might not feel comfortable about asking about their confusions about the work.

- Too often, these team members are supposed to work through their managers (even though the teams are supposed to use agile approaches) to resolve issues and questions. That means a question goes up the hierarchy, over, back down, and the answer makes the return trip. What is relevant is the communication directness and bandwidth.

To manage these challenges, make sure you have a team with all the skills and capabilities it needs. In addition, see the ideas in *Create Your Collaborative Team Workspace* on page 111 and *Build Respect with Working Agreements* on page 147.

4.6 Now Try This

1. Consider how your team type contributes to or detracts from the ability of the people to be a team. In the first chapter, we described an agile team. Do you have a team or a group of people? If a group, what could you do to create a team? (We'll discuss this in *Adapt Practices for Distributed Agile Teams* on page 147.)

2. Draw a picture of your team to learn your team type. (If you have multiple teams, maybe ask the teams to each draw their own image, so everyone understands what kind of a team they have.)

3. Does your team fit the business and project needs? Can your team members affiliate to deliver? Are the people able to communicate so they can collaborate?

Now that you know the kind of team you have, we'll talk about how teams can use various communication possibilities to collaborate.

CHAPTER 5

Communicate to Collaborate

Collocated teams use serendipity to collaborate—people eat lunch or get coffee at the same time and discuss issues in the work. Collocated team members might also take advantage of osmotic communication—things they can easily overhear or turn around to see—to clarify what the team is doing.

Distributed teams don't have the advantage of serendipity or of osmotic communication. Few distributed teams have the ability to easily leverage serendipity for informal meetings. They may not have the ability to meet in the hallway, in the coffee room, or a lunch room. And, if a team wants the benefit of osmotic communication, they need a dedicated team backchannel.

Distributed teams must purposefully make time for both formal and informal communications.

Informal approaches with our colleagues allow us to share insights, updates on our work, and even personal updates. (See *Create Resilience with a Holistic Culture* on page 33.) Distributed teams might need more formal approaches for team-based meetings, such as retrospectives, creating working agreements, and the various planning meetings.

Different distributed team types need different collaboration and communication mechanisms, depending on the intent of the communication as well as the hours of overlap.

The first question we hear about distributed teams is, "Which tools are best?" While many people mean "agile" tools, we also hear this question about communication tools. We won't specify any tools here, but we will suggest guidelines for selecting communication tools. We'll address the idea of agile tools in *Create Your Collaborative Team Workspace* on page 111.

Before we address how to communicate, let's first discuss if people feel safe to communicate.

5.1 Create Psychological Safety in Your Team

When teams create psychological safety, everyone feels free to raise issues and to fully participate. If people don't feel safe, they might not even use the tools available to them. Amy C. Edmondson, in *Teaming: How Organizations Learn, Innovate, and Compete in the Knowledge Economy* EDM12, discusses the need for psychological safety in interdependent collaborative teams. Safety allows the team members to discuss, explore together, and learn. As a leader, you might want to check out a tool for fostering psychological safety.[1]

Edmondson's five points about safety are:

- Use clear and direct language.
- Encourage learning from small experiments.
- Admit when we don't know.
- Acknowledge when we fail.
- Set boundaries for what is a personal or team decision and what is not.

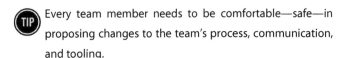 Every team member needs to be comfortable—safe—in proposing changes to the team's process, communication, and tooling.

If someone in a "leader" role takes responsibility for all innovation, then the group will always defer to that leader for ideas. This deference just becomes another form of waste on the team as the team will always be waiting on this individual. If the leader can make it safe to raise new ideas (no matter how outlandish those ideas appear) and

---

[1] https://rework.withgoogle.com/guides/understanding-team-effectiveness/steps/foster-psychological-safety/

help facilitate the evaluation of these ideas together, you now have an entire team focused on innovation.

Safety allows experimentation. When people feel safe, they often are willing to experiment. This could be individual experiments to learn new skills or explore some ideas which will later be brought back to the team. The more opportunities that can be provided for experimentation and the more the team is aligned on their overall vision (see *Define the Project Charter* on page 158), the more likely the team will take on larger experiments and produce significant innovations. Such leapfrog innovations might be key differentiators in the marketplace.

Leaders can encourage psychological safety. However, creating safety in the team is every person's job. When people feel safe, they are ready to speak up, which is especially key in distributed teams:

- When we frame the work as a learning problem instead of a problem of execution, we create the rationale for speaking up.
- When we acknowledge our own fallibility, we create safety for speaking up.
- When we model curiosity and ask lots of questions, we create a necessity for speaking up.

When team members have the rationale, the safety, and the necessity for speaking up, they can work through the problems they encounter. When they don't have the safety to speak, they often can't deliver the outcome the organization wants.

Without safety, team members may miss opportunities to explain key observations, or they may feel inhibited in reacting to possible severe implications for the business.

Safety allows us to take the risk and explain the situation, the learning, and the risks to others. Not being able to explain these risks have costs.

5.1.1 Lack of Psychological Safety has a Price

One distributed team had trouble discussing the product's architecture. The original architect, back at headquarters, many time zones removed from the team, had "decreed" this was the correct architecture. By the time this team tried to fit their new features into the architecture, no one could see how this architecture was useful. Because the team members didn't feel safe confronting the original architect, it took them months to add a feature that should have taken a week. The lack of safety created a huge cost for the feature and a huge delay.

Too often, the organization creates double binds for teams and managers. Teams *must* use the designated communication and workspace tools. Managers *must* use those tools to report into the governance mandates. And, teams often contend with already-established pricing for communication and workspace tools.

When the team can choose and configure its process, communication style, and tooling, the team members feel safer to experiment and innovate in that space. We've heard of too many teams going "around" the designated process, not abiding by the team's communication agreements, and not using the team tools.

Teams have a variety of low risk decisions: the dedicated team backchannel, the kind of board, and how to track the work on that board. When someone in power removes these decisions from the team, changing communication, process, and tooling configuration moves from low risk to high risk for the team. Increasing risks lowers psychological safety and creates increased risk for the team's communication, process improvements, and innovation.

5.1.2 Model Collaboration Through Safety

If people feel safe in their work environment, they are willing to contribute in more and different ways to their team. When psychological safety is high, people feel they can raise concerns, raise alternative opinions or ideas, and openly debate the merits of an idea. They are willing to learn and experiment more.

Brad, a coach, modeled safe behavior for a new team.

Encourage Safety Even in "Unsafe" Environments

Brad, a new-to-the-team agile coach for a newly formed distributed team, was concerned by his team's apparent unwillingness to collaborate. The team members were all in the same time zone (but different cities), and they were dispersed, all working from their homes as a nebula team. When he looked at their board, every person had their own story and there were three code reviews and two test reviews in the waiting-for-review column.

At first glance, he wasn't sure this was a safety issue. As he listened to their conversations, he wondered if he was missing something. He conducted several one-on-ones with each person to understand their concerns. In the one-on-ones, he planted the seeds of safety ideas and asked for help.

Each person was concerned that they couldn't "fail" in any number of ways: What if their code wasn't elegant enough? What if their tests didn't find everything? What if the product performance wasn't good enough? Each person felt as if *they* would fail, not their work product.

Brad invited the team to a "learning" meeting. In the invitation he said, "I need to learn the ways you would like me to coach you." In the meeting, he said, "I don't know everything I need to know to help you immediately. I would like your help in discussing when I make mistakes, when I don't know the boundaries for my role, and how to help the team learn from small experiments. I would like to learn early, not fail at my role."

The team members were surprised. One of them, a senior developer, said, "Since we're all new on this project, why don't we do the same thing?"

By modeling safety-building behavior, Brad helped the team learn how to create safety in the team and work together.

Brad used the ideas in *Asking for Help Can Build Respect* on page 150 to help build safety in the team.

If people feel safe about raising issues, concerns, and ideas, they will raise concerns when they see something that may cause harm to the team, the business, or their customers.

When team members don't feel safe, they don't raise those issues. They assume the "leader" will take responsibility for problems and risks. This may be a manager, a Product Owner, a Scrum Master, a coach, or even a senior team member who takes on a role of "protector" for the team. However, this means that you have fewer eyes looking for project risk.

Without psychological safety, team members are less willing to experiment and the work of the team becomes "business as usual." Some may be fine working this way, but your market may not be, as competitors innovate beyond your capabilities.

When teams feel safe, they learn together. They communicate, they create their collaborative team workspace together, and ship products on a regular basis.

Now we can see how various communication approaches fit some teams better than others, and review some considerations for your team's communication.

5.2 Use the Appropriate Communication Channels

Distributed teams have several possibilities for rich and natural communication.

5.2.1 Understand Communication Richness and Naturalness

If you've read any of the research about communication at distance, you may have heard of "media richness" theory. Media richness, as described in Daft and Lengel's *Organizational Information*

Requirements, Media Richness and Structural Design in *Management Science*, DAL86, solves two problems: the absence of information and the ambiguity of information.

Media richness theory explains why people choose various media, such as email, voice, or video. People select what they think is the the most "appropriate" communication channel to learn what they need to learn and to resolve ambiguity.

Daft and Lengel define media richness in these ways:

- Ability to handle multiple information cues simultaneously.
- Ability to facilitate rapid feedback.
- Ability to establish a personal focus.
- Ability to utilize natural language.

The richer the medium, the easier it is to understand the information and refine our understanding.

There is a competing theory called media naturalness described in *Media Richness or Media Naturalness?*, (KOC05). Kock argues that face-to-face is the communication channel likely to lead to the least cognitive effort, the least ambiguity, and the most personal engagement in the communication. The closer we are to face-to-face communication, the easier we find our communication.

Media naturalness allows us to see and hear each other as humans, with all the other cues of our communication: grins, frowns, sarcasm, and sincerity.

Media naturalness includes:

- A shared environment which allows people to see and hear each other.
- A high degree of synchronicity to quickly exchange ideas.
- The ability to convey and observe facial expressions.
- The ability to convey and observe body language.
- The ability to convey and listen to speech.

Both of these theories, rich and natural, are hypotheses. We will use the term rich and natural in this book. We are not taking sides.

However, we have observed teams who use both rich and natural communication channels tend to be more effective.

 Choose rich and natural communication as often as possible.

Distributed teams can create a rich and natural communication environment with judicious selection of tools. Don't assume one tool will fit all your team's needs.

Instead, consider using a set of tools that support rich and natural communication.

Humans Need Rich and Natural Communication

When we wrote this book, we used a synchronous audio/video tool (Zoom) along with email and cloud storage for keeping notes. Zoom allows us to not just hear each other, but also see each other's faces. We can then notice the human cues of what we say.

Johanna tends to be sarcastic and an eye-roller. When Mark can see Johanna roll her eyes and gesticulate wildly at the screen, he knows he can enjoy the moment, wait for her rants to settle, and then continue the work.

When Johanna sees Mark rest his fingers on his chin, she knows he is thinking. She knows to stay quiet so he can think and explain where he wants to go. Rich and natural communications enhanced our collaboration.

If the distributed team members only have access to asynchronous communication, they can't gather cues like these. In our experience, distributed teams who use natural and rich synchronous communication tools can better understand each other, make allowances for personal idiosyncrasies, and help people build their success. The richer and more natural the communications, the more the team can realize the principles described throughout this book.

5.2.2 *But We Can't All Use Video!*

Not all distributed team members can use video. They might have restricted network bandwidth in their location. They may be working late in the evening and need to avoid disturbing others. Or, they may have concerns about having a camera on if they are working in their home.

In such scenarios, teams can still use rich communication channels that allow them to coordinate and collaborate on the work. However, these tools may seem less "natural" without video for face-to-face communication.

In that case, we recommend that teams should be allowed to collaboratively select the technology that works best for them and create team working agreements. As team members leave or join the team, the team can revisit their choices and agreements.

5.2.3 *Not All "Synchronous" Communication Channels Are Full-Duplex*

Both of us have worked in organizations where we had access to supposedly synchronous speakerphone systems. These systems were not full-duplex; each side could not talk at the same time. Instead, one side could talk and the other would listen. If the person on side A spoke first and continued to speak, side A "won" and side B could—literally— not get a word in.

If your organization has one of these speaker phone systems, do not fool yourself into thinking you have rich, full-duplex communication. You do not. You have fast asynchronous communication.

You can use a chat backchannel to overcome some of this communication lag. And you might also consider how to replace the half-duplex speakerphone system.

5.3 See Your Team's Communication Options

Teams have choices of which communication channel to use, depending on the hours of overlap and need for learning and feedback. The key is for teams to choose when to balance the synchronous and asynchronous communication; to balance the deep focus individual work with the deep focus small group work, to a broader focus of collaboration with their entire team or (possibly) multiple distributed teams.

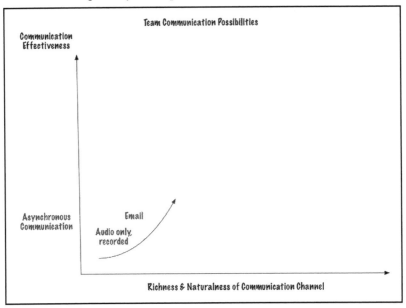

Figure 5.1: Asynchronous Communication Channels

Any time you do not have rapid feedback with the other person, you have asynchronous communication.

Asynchronous communication can be documentation, recorded voicemails, or email. You might even leave messages for each other on real or virtual whiteboards. The problem with asynchronous communication is that it is low bandwidth. People can easily misunderstand each other in asynchronous communication.

When teams have no hours of overlap, they can only use asynchronous communication, the *least* rich and natural communication channels. That creates several problems:

- Each of us assumes everyone receives the communication when we send it. However, botched recordings or problems with email can delay or prevent reception. Our email systems don't always send email when we expect them to do so. If the conversation continues, the context changes. The communication might no longer make any sense, leading to more confusion and possible delays as each person explains their current context.

- The least rich communication channels invite misunderstanding. Asynchronous communication can cause erosion of trust. Team members might even wonder if the other person is working. We jump to assumptions about others when we have no or delayed feedback.

- The team's cycle time increases because it takes more time than anyone expects to ask and answer questions. Team members might not realize the urgency of the question or the need for the answer.

Synchronous communication creates more possibility for understanding because people can share their context.

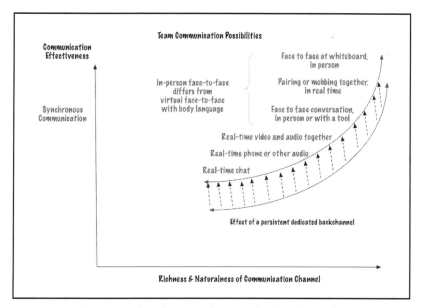

Figure 5.2: Synchronous Communication Channels

Synchronous communication allow teams to ask and answer questions much more quickly. This is one of the reasons hours of overlap matter to all distributed teams.

Teams can use video to understand each other and work together in real time. These tools are almost as rich as being in person in the same room. Video helps us understand each other's context. With rich and natural communications, teams can pair and mob together, even if they are not collocated.

Real-time, in-person work at a whiteboard is the most rich and natural communications channel. We don't see *distributed* teams using this form of communication well or as frequently as they need. That's one of the reasons we recommend teams get together in person on a regular basis.

We have noticed that teams who use a dedicated backchannel simultaneously with rich communication channels can improve the effectiveness of their communication. During online meetings, the team members use the active dedicated backchannels to help answer

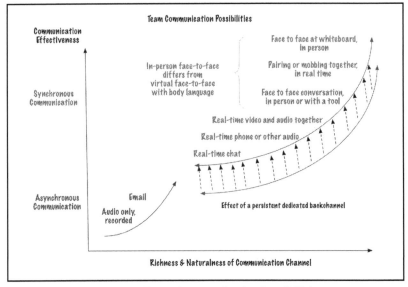

Figure 5.3: Team Communication Possibilities

questions. Sometimes, they use the backchannel to provide additional context for new team members or for visitors who are not part of the team's day-to-day work.

Consider these communication options depending on your team's hours of overlap.

Hours of Overlap	Default Communication Channel	Secondary communication channel, open and public backchannels
4-8 hours (half to a full day of overlap)	Synchronous, such as video, audio, chat. Notice that distributed teams, at this writing, cannot easily draw together, as at a whiteboard.	Chat, text, and email as a backup for synchronous communication. During hours of overlap, the backchannel is supplemental and can help define the problems the team works on.
1-4 hours (Up to a half-day of overlap)	Some synchronous during the hours of overlap, which includes audio, video and chat.	During hours of overlap, chat, text, and email as a backup for synchronous communication. Outside of hours of overlap, only asynchronous channels.
Zero hours of overlap	All asynchronous communication, such as email and possibly chat.	Only asynchronous channels. No backchannel unless people time-shift to create hours of overlap.

Figure 5.4: Communications Options for a Distributed Team Based on Hours of Overlap

Teams with too few hours of overlap are at a disadvantage for their communication supporting their collaborative and innovative work. See the chapter *Avoid Chaos with Insufficient Hours of Overlap* on page 41 for a fuller discussion.

5.4 Build Consensus for Team Communication Preferences

Once the team agrees to its communication preferences, work with each person on the team to adopt the team's preferences.

One team had problems with how one member preferred to work. He preferred a separate board for his work, not using the team's board. He preferred a non-email application for the kinds of communication other people used email for.

His preferences affected everyone's communication in the first iteration. He was unable to make the retrospective, so one team member offered feedback. He acknowledged the feedback and chose not to change.

In the second iteration, the communication problems increased. At this point, several people tried to schedule one-on-ones to discuss these problems. However, he had no more than two hours of overlap during normal work hours, and he met with the organization's customers during that time.

By the end of the second iteration, the team was ready to kick him out. The team leader finally was able to make time to discuss these problems with him.

The problems never quite got resolved. He didn't feel as if he was a "real" member of the team, and the other team members didn't respect him enough. At the end of the project, he left that team and moved to a team with more hours of overlap.

Early communication problems don't vanish. They increase.

5.5 Enhance Discussions with Dedicated Backchannels

We encourage teams to use *dedicated* backchannels: a designated chat channel to support video conferencing or any synchronous or asynchronous conversations. Many teams use a dedicated backchannel to create a distributed form of osmotic communication.

The dedicated backchannel is the default for the team and is always available—it adds context and allows people to collaborate on the work. In addition, all the team and only the team uses this backchannel.

Imagine you are new to a distributed software team. You don't know the product. You're not even sure you know what the customers want. You're working on a new feature with a colleague, and that colleague has been a part of the product team since its inception, five years ago. The colleague is not in your time zone.

The colleague has the entire context of this feature. You don't have the context. You might feel vulnerable asking a question about the context..

With a dedicated backchannel, the team should *expect* and encourage questions like this. You are probably not alone in your question, and the entire team can benefit from your colleague's explanation of the entire feature context.

That explanation begins to establish psychological safety with the new team member and encourage others to share information. (See *Create Psychological Safety in Your Team* on page 88 for more details.) In addition, the information can help teams identify "problem" areas in the product.

5.5.1 Why Use a Default and Dedicated Chat Backchannel

You might wonder why we don't immediately recommend sending the new team member to internal documentation. Neither of us has seen internal documentation stay current over several releases. If your team can maintain that internal documentation, terrific. Especially with distributed teams, we have had more success with explanations at the necessary time. However, there is still a place for internal documentation.

As a team shares information in the backchannel, the team may encourage a new team member to capture a summary in online documentation or a team repository. Now the new team member focuses on a specific area of the documentation to explore rather than trying to absorb all of the documentation. This may generate more focused questions from the new team member to deepen their learning.

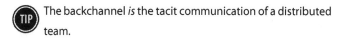 The backchannel *is* the tacit communication of a distributed team.

Make sure the default and dedicated backchannel is available to all the team members 24 hours a day, regardless of any other communication tools. We refer to this as an "always accessible" communications channel.

Here's an example. We used the Zoom meeting tool to write this book, because we like the video and audio capabilities of the tool. We did not often use chat in Zoom. Not because it's not helpful, but because it's not *persistent* for us to keep the context of our discussion. Even though Zoom retains the chat as a saved file when the conversation ends, we would have to look outside our team space for the information.

When teams use a tool-based chat, the chat is gone when the tool stops. Instead, we recommend the team use the team's default chat channel, not a meeting tool's chat channel.

Distributed teams may use a backchannel in place of meetings. Great meetings have agendas and minutes, so everyone can see why they made decisions.

Distributed teams with insufficient hours of overlap might decide to not meet. Instead of a meeting, they can use the dedicated backchannel to see the context of their discussions, their reasoning about decisions, and the decisions themselves.

These discussions reinforce your team's culture. If your team needs to change its chat tool, see if there is a way to migrate the searchable history to the new tool. That way, the team doesn't lose the prior discussions with all the context.

5.5.2 Benefits of a Dedicated Backchannel

When teams use dedicated team backchannels, the team members benefit in several ways. First, when everyone asks and answers in their dedicated channel, the team learns together. They explore knowledge gaps and answers for everyone, not just the person who asked the question.

We assume that this dedicated backchannel is open to the organization, and dedicated to this team's use. We do not encourage private backchannels for significant conversations about the work.

Private backchannels can reinforce a "them" vs."us" approach. And, the team loses the insights other people might contribute. Private channels may make it easier to avoid the difficult, but necessary conversations the team needs to resolve the difficult problems. The loss of insights plus the addition of gossip and insufficient safety make private communication challenging for distributed teams.

The team can use the dedicated backchannel for clarification purposes:

- When people have non-urgent questions.
- When people have questions they think have simple answers.
- When team members want to draw others' attention to something in the product, as in, "Did you see this performance change?" or "Did you notice these log messages?"

Encourage your team to use dedicated backchannels to check on each other. And, encourage your team to use the richest possible communication channels when they have trouble understanding each other or agreeing on an action.

See *Create Psychological Safety in Your Team* on page 88 for a discussion of how team safety affects the team's ability to use any sort of communication channel.

5.5.3 Backchannels Don't Work for Everyone

In collocated teams, people are wired to use multiple channels of input. We do this without thinking. Some of us are better than others at managing the variety of input.

On distributed teams, people are more likely to restrict the variety of channels of input.

The backchannel creates one more online channel for people to juggle. Some people don't find that juggling helpful. For example, if they use video and screen sharing to mob on the work, they might not want to check the backchannel.

If your team has someone like that, honor their preferences. You might ask a facilitator to verbalize what's occurring on the backchannel so the person can stay fully engaged.

When people collaborate closely with colleagues, focusing on the work, the backchannel can distract more than aid the work. In such cases, the team should discuss working agreements about how long an individual, pair, or mob might stay in focus and when they need to check in with the other team members.

5.6 Language Matters

Our language reveals our assumptions and our mindset.

We've seen several communication/collaboration problems with distributed teams. Here are three language problems that appear to prevent easy collaboration in distributed teams:

- Differentiating by "headquarters" or "remote" or "offshore"
- Naming some time zones "sleepy" or "early"
- Using absolute or relative language

5.6.1 *Headquarters, Remote, Offshore*

Naming teams "Headquarters" and "Remote" or "Offshore" implies a hierarchy. You might not think a Headquarters team implies a hierarchy but, too often, it does. And, too often, "Remote" and "Offshore" mean "second-class." And, when the Offshore people are contractors, the organization might have a culture of considering them third or fourth class.

We recommend naming teams by their product, not by their location. The side benefit of naming teams by their product is it helps everyone realize that the team affiliates around the product, not other work the manager might want the team member to do.

Some teams prefer a totally agnostic name, such as the Purple Team, where the team had a day every week when everyone wore purple. Or the Camo team, where everyone wore a camo cap when they were on video.

Also, by naming the team by product or an agnostic name, you do not need to change the team name if you change team members who work in different time zones.

Sometimes, location *plus* the product name works. Recall the East and West Payments teams from *Shift to a Mindset of Collaboration* on page 9. The managers saw one more effect of naming the teams "East" and "West." When they wanted to add more people to one or the other team, they actually thought about where the new team members were located, to create the maximum hours of overlap for the team. This approach works as it does not indicate a status or hierarchy.

5.6.2 *Sleepy versus Early*

We've seen some teams whose members are across more than three time zones and refer to their Western teammates as "sleepy" or Eastern teammates as "early." While you might not think the sleepy or early is a denigrating term, we have seen these terms used to bully people.

No one lives on a "sleepy" coast. No one lives in an "early" time zone. They live where they live.

Teams who denigrate each other's location in any way don't tend to collaborate well.

5.6.3 *Relative versus Absolute Language for Measurements*

We like to compare temperature (especially when it's winter) when we work with people across the globe. Sometimes, we envy their temperatures, and sometimes we don't. It may also be a good reminder of which hemisphere our colleagues work in (e.g., teammates in the southern hemisphere will have the opposite seasons to their colleagues in the northern hemisphere).

When we talk about "31 degrees," is that Fahrenheit or Celsius? 31 creates different feelings in the two temperature scales.

When we talk about "top of the hour" instead of 3 P.M. (12-hour) or 1500 (24-hour), we allow everyone to fully participate. Each of us has a preference for how we designate time. We recommend you use 24-hour time for meetings across the globe, and ask people if they are available until the "top of the hour" or "half past the hour."

These small changes in language, moving from absolute to relative, helps everyone communicate better.

5.7 See Your Team's Communication Traps

We've seen several distributed team communication traps:

- Teams treat the dedicated backchannel as any other communication channel.
- Team members default to one-to-one communication.
- Some team members are video-averse.
- Some team members default to asynchronous communications for everything.
- People focus on only one communication channel at a time.

5.7.1 Trap: Treat the Dedicated Backchannel as Any Other Communication Channel

The dedicated backchannel for distributed teams is a vital part of the distributed team workspace. You cannot effectively work with the team if you are not using that dedicated backchannel.

Compare the dedicated backchannel to walking onto the shop floor. If you don't enter the team workspace, you will not be able to see what the team is working on, hear the problems they wrestle with, or have an easy way to ask them questions about the work. (See *Create Your Collaborative Team Workspace* on page 111 for more details.)

In addition to the team's dedicated backchannel, establish other chat channels for cross-team discussions and socialization.

5.7.2 Trap: Team Members Default to One-on-One Communication

Some team members, accustomed to using a one-on-one private backchannel on collocated teams, continue to use that private backchannel, instead of the dedicated backchannel for team-based communication.

This trap might be an indication that your team members don't feel safe enough to communicate openly on some topics with the entire team.

Private backchannels default to non-collaboration, rather than team collaboration. It's possible a pair might check in with each other, "Did you mean to say that?" to verify the communication. However, we have seen more instances of clique-building, rather than supportive team discussions. (For more, see *Create Transparency at All Levels* on page 25.)

We have seen private backchannels work effectively if a team member has an urgent issue and lets one person on the team know. One parent texted "Need to take my son to the hospital!" to just one team member. That was fine because the team member let the rest of the team know. The recipient can then field the team's responses, so the parent with the emergency doesn't see all the other messages.

5.7.3 Trap: Team Members are Video-Averse

The more collaborative a team needs to be, the more they need to use video. And, not all team members are comfortable with video.

Sometimes, this trap indicates that the team member doesn't feel safe enough to use video.

When team members are not so happy with video, don't force them. Model the video-on behavior, but never force other people on your team to join you. Also, ask them privately what makes them uncomfortable about the camera. Some people have moved the orientation of their desk so their house doesn't show in the camera's field of view.

We've said before that we don't recommend mandating video for team members. The retrospective is the one exception. Without video, you can't see what people are happy about or uncomfortable with. You can't tell. See *Use Video for Your Retrospectives* on page 175.

5.7.4 Trap: Default to Asynchronous Communications

Especially when teams have few to no hours of overlap, they may feel as if they have no opportunity for synchronous communications. Such teams rely on asynchronous communications for all their work.

We have not seen asynchronous-only communications be successful in the long-term, especially for agile teams. We have seen some asynchronous communications work, especially with handoffs.

If your team has few to no hours of overlap, consider the following options:

1. Synchronize your team at least once a quarter—and preferably more often—with an in-person meeting.
2. Ask people to time-shift—once in a while, and not the same people all the time—so the team can take advantage of richer communication modes.
3. See if your team can recreate itself to have more hours of overlap. Some possibilities are people permanently time-shifting, people moving, and changing people on the team so the team members have more hours of overlap.

We have seen that teams who primarily use asynchronous communications have many more misunderstandings than teams who synch on a regular basis.

5.7.5 Trap: Team Members Focus on Only One Communication Channel at a Time

Collocated team members might not realize they use multiple communication channels in meetings. Collocated team members use face-to-face communications, notes on a whiteboard, and possibly meeting notes in a team workspace.

Distributed team members can mimic all those communication channels. The face-to-face becomes real-time video and audio. The notes on a whiteboard might be in the dedicated backchannel. And the meeting notes in a team workspace might be the backchannel or some other note-taking location.

When teams use all their communication channels, they learn the team's context. They realize how to coordinate the team's work. The all learn more about the team's tools and product.

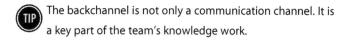

> The backchannel is not only a communication channel. It is a key part of the team's knowledge work.

Some team members may be uncomfortable with all the active communication channels. How can they "focus" on audio, video, and chat?

- Can people learn to be comfortable with multiple channels?
- Can these people manage the information flow in other ways?
- If they focus on one channel, can they still understand the entire context of the meeting?

People new to distributed work might not realize how easy it is to engage in all three channels at the same time with practice. Encourage people to practice and know that while they practice they might miss information.

Sometimes, people don't have enough screen real estate to easily manage the various channels. Make sure that the person has sufficient screen space for audio, video, and chat.

Successful distributed agile team members have the mindset of collaboration, experimentation, and principles over practices. If team members are not able to sufficiently collaborate, experiment, or use the principles of multiple communication channels, are they suited for a distributed agile team?

We can't answer that question for you. Every person and every team is different. We caution you about teams whose members cannot use all the communication channels available.

5.8 Now Try This

1. Assess your team's hours of overlap to understand the communication channel richness and naturalness they can expect now. Does your team have sufficient communication channels? If not, what can you do?

2. List all the different channels your team uses now to communicate. How many of those channels are dedicated to the team? Does your team have one persistent place they can look to establish and search context (i.e., a dedicated backchannel)? Does the team publicly ask and answer questions in this dedicated team backchannel? For teammates? For those outside the team?

3. Assess your team's communications capabilities. Does the team have the capability to use real-time video and audio? If not, what would you need to do to help the team create a natural communication environment? Is anyone on the team video-adverse? How can you help them feel comfortable with video? (We have hints in the next chapter on team workspace.)

Now that you've seen the communication options for your team, we'll discuss how to create your team's virtual workspace.

CHAPTER 6
Create Your Collaborative Team Workspace

Collocated agile teams share a real workspace: the team room, people's offices, and anywhere the team meets to work or talk about the work. Their physical workspace helps them create the human connections they need to complete the work.

All agile teams use some form of a board to plan, track, and reflect on the work. Many collocated teams use a physical board. Collocated teams might not realize all the various places they use for the team's workspace.

Distributed teams also need a workspace to create human connection and to visualize their work. Distributed teams intentionally create their team workspaces, for their teamwork and project history.

An agile team's first decision that affects their tooling is a board that reflects the state of the work. For many current tools, that often means deciding whether they should use iterations or flow for their approach.

6.1 Select Iterations or Flow

Collocated agile teams often start with an iteration-based approach, such as Scrum, for their work: they plan for an iteration, they have standups each day, they refine stories in preparation for the next iteration, and they demonstrate and have an iteration review at the end of the iteration.

If your team doesn't have at least four hours of overlap, or if your team is not accustomed to collaborating on the planning, work, and reflection, or if your team is supposed to work on multiple products during an iteration, we advise against an iteration-based approach.

We have seen more teams succeed with a flow-based agile approach, such as kanban or lean development. See *Create Your Successful Agile Project*, ROT17, for more specifics on agile approaches.

Agile practices enable a team to collaborate to produce value early and often. What kinds of agile practices does your team need? Let's start with the meetings.

- Does your team need a cadence or a timebox for their planning, demos, and reviews?
- Do they need to reconnect on the work every day, or do they hand off?
- Do they need to collaborate more intensely via pairing or mobbing on a solution?

If your team works on several products over a two-week time period, or has insufficient hours of overlap for collaboration, consider flow-based agile approaches.

First, let's discuss the difference between an iteration of work and a cadence for the work.

Iterations are timeboxes. Many agile teams select two weeks as their iteration duration. That's long enough to finish work and short enough that if you had to throw everything away, it would not kill the project.

The idea with a timebox is that you are done—by definition—at the end of that timebox. You don't extend the timebox. Neither do you let remaining work slide into the next iteration.

When teams use a cadence, they decide when to plan, review, and reflect on a regular basis. However, each of these meetings may be on a different cadence based on what the team needs to maintain a flow of work. Some teams workshop stories every Tuesday morning. Some teams review their work every Wednesday afternoon. Some teams select Monday afternoons for their kaizen/reflection. It depends on the hours of overlap and when people are available.

Iterations require many hours of overlap to plan, collaborate, and complete the work. Teams with few hours of overlap might be able to use iterations. These teams often discover that they have trouble defining when the iteration starts/ends and difficulty finding sufficient hours for collaboration.

If your team has fewer hours of overlap, we recommend the team use a kanban board to visualize all the work and the flow of the work. To expose and eliminate bottlenecks, the team can use a cadence of planning, demos, and review.

6.2 Help Your Team Visualize Their Work with a Board

If your distributed team is new to agile approaches, you might think the team needs an electronic board in a tool to start. We don't recommend that.

Teams who are new to agile approaches and teams who are new to working together need to experiment. We find a physical board best supports experimentation.

A new team met each other for the first time in their agile training. They had all traveled to the company's headquarters to meet each other and learn about agile in the form of Scrum. They had two days of training, and then another three days to work together to understand their project vision, release criteria, and to start on their backlog.

The team tried a typical physical Scrum board with three columns: Ready, In Progress, and Done. The team worked together, as a collocated team for the next three days. When they worked together, the board was fine.

The team members returned to their offices to continue work for

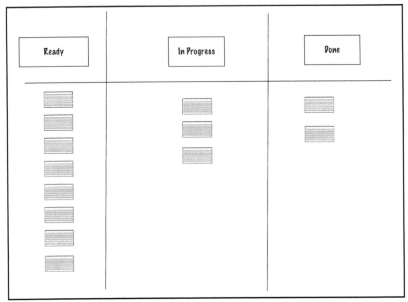

Figure 6.1: Scrum Board

the next week. However, at their first retrospective, they realized they had too much work In Progress, not Done.

Brad, an experienced agile project manager, was their Scrum Master. He recognized the team had work in progress in several areas. He asked the team if they would be willing to try more columns on the board.

The team modified their board to see where the work was. That board looked more like a serial project, with these columns: In Analysis, Code and Unit Test, In Test, Waiting for Approval, and Done. See Figure 6.2.

The team members were worried—were they really being agile? Brad suggested that until they could see *where* the work was, they couldn't know.

Brad added index cards to the relevant columns and took a picture of the board each morning and posted the image to the team's virtual workspace, a wiki.

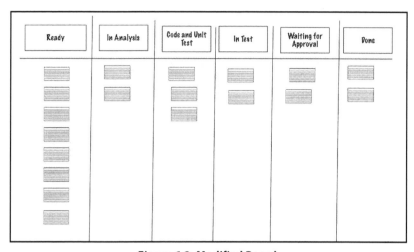

Figure 6.2: Modified Board

As the team worked through their next three iterations, they learned where their work was. Sometimes, work got stuck In Analysis because the stories were too large. Sometimes, the work got stuck In Test because they had one tester who was sometimes overwhelmed with too many simultaneous finished Code and Unit Test stories. That tester was the only person testing the completed Code and Unit Test work.

The physical board allowed them to experiment with WIP limits and treat their board like a kanban. This also allowed the team to move the stories across the board, not the people moving tasks across the board. Brad was in charge of moving the cards and updating the wiki with new images.

They discovered they rarely moved more than one card a day. Brad continued to take a picture of the board each morning. Team members used the dedicated team backchannel to discuss where their items were.

As the team changed their WIP limits and understood how to work well together, they ended up with these columns: Stories to Workshop, Ready, Dev/Unit Test, System Test, Accept, and Done. They had WIP limits on each column. Those changes helped the team work together and create a more predictable throughput.

The team had a subtle change in how they approached the stories. The team decided the stories were "Ready" *after* the team workshopped the stories. That allowed the developers and testers to perform a little analysis in the workshopping. The Dev/Unit test column in the second board encouraged the developers and testers to work together to complete analysis if necessary, and provide the tester with enough information to create system-level tests.

To the casual observer, the Dev/Unit test column looks like a coding activity. It is not. It is a team activity that helps the entire team understand the details of the intent of the story so everyone can create relevant code and tests for this one story. The team members can look ahead and say, "This might be useful for the next story we workshop." However, the team members manage their WIP for these stories and for the future.

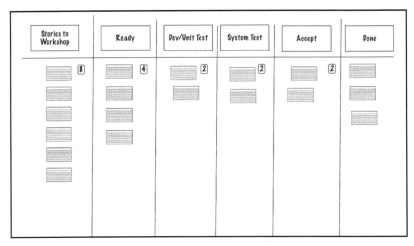

Figure 6.3: Updated Kanban Board

Once this team understood how to use agile approaches and how to use their board, they moved to a virtual board.

If your team has matured in their agile approaches and is distributed, they can still make changes in their board.

 Help your team practice their agility and design their board to signal blocked or waiting work.

One team was responsible for integrating several products together. This integration included internally developed products and partner products. Occasionally, this team noticed that the flow of work had shifted based on the types of integration work. They used their retrospectives to talk about what the new flow might look like and how they would track it on their board.

Sometimes there would be more manual testing and they would want to watch that more closely. Other times, the products being integrated would have a robust suite of automated tests. In this situation, the team would combine some of the review and test stages of their board. The team sometimes made these changes by "sketching them out first" and other times they would change the board together in their online project management tool.

Another team wanted to try kanban with a cadence. The team learned about kanban and decided to specify their WIP limits. They decided to keep the board as simple as possible and see what the board told them. They developed a board in Google sheets to see what worked for them. After two iterations of their sheets-based board, they then recreated the board in an agile planning tool.

6.3 Help Your Team Create a Board That Fits Their Needs

Often, distributed teams start with one agile approach and realize they need to integrate another agile approach.

Don't start with an agile project management tool or with someone else's recommendation for a tool. What works for one team might not work for another. Someone else's tool might restrict how your team thinks and acts.

Instead, start with your team's agile approach. Model that approach in a physical board or, if necessary, an electronic tool such as a spreadsheet, so the team can refine their approach before they move to a tool.

Consider these tips to help your team create the board it needs:

- Educate the team on their options for the kind of agile approach that will work best.
- Start with paper or a shareable document of some sort (such as a doc, spreadsheet, drawing) if possible. We have had good results with an always-on camera and one person whose job is to move cards on the board. If you prefer an electronic board, start with a publicly accessible spreadsheet for ease of collaboration and use. The team can discuss the columns they need, whether they want WIP limits on any column, and what those limits are.
- Experiment with the board, especially if your team is new to agile approaches, so the team can gain experience with this particular board before the team moves to a tool. Consider the daily pictures, a camera on the board, or some other way to see and move cards on a regular basis. Beware of moving to a tool too fast.
- When the team does move to a tool, can the team make the changes to the board or do they need to wait for an administrator? We recommend giving the team control of their own board so they can quickly discuss and make small changes.

Each team is unique. Each team has more or less ease with agile approaches. Help your team decide on a board that exposes their system of work—their current context—not someone else's idea of what a board should be.

Many distributed team members struggle with work that pulls them away from the team. You can help your team stay focused on the team's work.

6.4 Identify Your Team's Focus

Non-collocated teams often have problems with the focus of their work. Too often, the team members are supposed to "focus" on work for their functional manager, not the team's work.

Here is a possible activity for your team:

1. Ask each team member to identify all the work they are supposed to do. How much of their work helps their team achieve their goal versus which work helps their functional manager? (We often ask people to make a list on a legal pad and write "T" for the team's work and "M" for manager work.)
2. Ask the team—as a team—to develop at least three ideas to focus on their team's work.
3. See if those ideas reveal impediments or challenges that the team and/or the team's leadership can help address. In other words, what would make it difficult to follow through on the ideas?
4. Ask the team's leadership to address the problem of manager-based or functional work that does not help the team deliver its work. While we can't necessarily stop work immediately, we can reduce it or eliminate it over time, as long as the managers realize the team members need to focus on their work.

The goal of an agile team is to deliver the team's work, as a team. While the managers might also have work they want the team to accomplish, the managers need to work with the relevant product owners/customers/responsible person, so that the manager can assign that work to some team. However, this team might not deliver that manager's work because that work is not in the team's context.

6.5 Distributed Teams Create Their Own Context

Collocated teams share a physical space, which includes their board and their entire project context. Collocated teams might not have to be explicit about creating their workspace.

While distributed team members share the same project intent, the members don't share the same workspace. That means each distributed team needs to create its own shared virtual context—the system of the team's work and their team's workspace.

Creating that context helps the team learn how to work together and how to best use their team workspace. As the team learns to work together, they might realign this context during the project.

All agile teams use working agreements to create their project context. We'll discuss working agreements in more detail in *Build Respect with Working Agreements* on page 147.

A team's context contains the tools the team uses, who has access to the tools, and each person's individual setup.

First, is the team's tools requirements.

6.6 Consider Your Team's Tools Needs

Just as a team defines their context, every distributed agile team needs several tools to build personal relationships and deliver great products. Every team member needs the ability to initiate and use rich and natural communications as in *Use the Appropriate Communication Channels* on page 92.

However, before they can select tools, teams need psychological safety. Once they select tools, they need the safety to continue to use the tools successfully. This includes being comfortable in reconfiguring the tools and being aware of the impacts described previously.

While writing, we circled around the problem of do we recommend teams create psychological safety first or select tools first. If people don't feel safe to discuss their needs for tools, they get stuck with what the team—or more often, one person—picks. That person may not even be on the team. On the other hand, selecting tools that fit the team's needs can enhance psychological safety.

Aside from tools to build personal relationships and collaborate on the work, teams need access to various product development tools so they can deliver great products. These tools include ways to track their work, discuss their work, and archive knowledge. Distributed agile teams might discover that a collaborative IDE (Integrated Development Environment) is necessary for their ability to pair and mob.

How can your team create all-team access to at least these tools?

1. The code repository.
2. The test repository.
3. The defect repository. That repository might differ from the backlog.
4. The place to store backlog items for near term work on feature sets.
5. The place to see the roadmaps for the longer term. (In a program, teams might need to see other teams' roadmaps to gain perspective on upcoming work.) Roadmaps clarify the product strategy, defining the feature sets in general terms.
6. The team's project workspace so the team can put project documents in one specific place.
7. The place to gather team data. This includes a board and the team's metrics.
8. The place to report project status.

Your team might need different or additional tools. For example, consider your team's communication channels, both specific to the team and across the organization.

All teams require tools that fit their specific context. We'll start with the team's information radiator, the team board.

6.6.1 Teams Need Tools That Fit their Unique Context

Each team needs to visualize its own data because each team has different challenges.

One team started with a Scrum board in a well-known tool. However, they needed to use a kanban board to see their WIP and where the work was stuck.

Management wanted to be able to generate metrics from the team's board. While we recommend against management viewing team-level metrics, that is what this management team thought they wanted. The team needed different data than management thought they needed.

This team could not change their board configuration themselves. They needed Director-level intervention on behalf of the team for the administrator to change the board. The Scrum Master manually generated the project measures and then spent time to educate the managers, including the Director-level managers, as to why the measures the managers wanted didn't help and what the managers needed.

The team finally was able to see their WIP and where the work was. They realized they had a bottleneck in testing and were able to use a combination of more tests in development and more automation hooks for the tester. This team chose to continue with a kanban board. They kept the cadence of the iteration with planning and demos and retros, but the board didn't look like a typical Scrum board.

Teams might need different tools than managers think the teams need.

When teams don't have access to the tools they need, they may find a way to get access. For instance, some teams we know use a video communication tool that someone pays for out of their personal pocket, instead of the corporate communication tool.

One of the reasons these teams don't use the corporate communication tool is that not everyone has a license to that tool. Only certain people do.

6.6.2 *Shared Tool Access Creates Shared Responsibility for Context*

We have met many team members who do not have access to all the "team's" tools. These team members might not have sufficient Wi-Fi access. Or, the team doesn't have sufficient licenses for everyone to have access to all the tools.

Teams members—not the facilities or finance people—must choose and agree how to use their tools. The team does need to understand other people's context so they make good decisions. However, the team decides.

Many new-to-agile teams might want to restrict a new team member's access to particular areas of the code or applications until that new person gains specific experience. Until team members learn how to experiment safely, it might be okay to restrict access to some parts of the team's responsibility. We might suggest that teams mistake-proof their environments, but that takes time that teams might not feel as if they have. The real question is this: When does the team extend "privileges" to all the team members?

We have seen team members discuss access to applications and administrator-level system access based on other people's need to know. Avoid this need-to-know problem. That creates bottlenecks in the team. (See *Think in Flow Efficiency* on page 10.)

Here are some ways to test when the team member context might change:

- Does the proposed access affect the team's process? We are in favor of every team member being able to affect the team's process. For example, some teams choose to modify their core hours to accommodate a digital nomad's shift in time zones. When teams are willing, this is a terrific way to enhance teamwork.
- Does the proposed access affect other people inside the organization? What if the team wants to change a server location, do they need to let other people inside the organization know beforehand, and with how much notice? We have seen teams move the demo server and even change the demo version of the application with no notice to Support, Sales, or Marketing. Changing servers affected other groups, which created problems for the rest of the organization. New team members might not realize that changes to servers, applications, or even the look

and feel of the user interface might affect people inside the organization, not just outside the organization.

- Does the proposed access affect the product or people outside the organization? For example, can all team members change the build or install scripts, or do people need an "apprenticeship" period? Do all team members rotate through this apprenticeship? If not, why not? We have found that when team members understand what it takes to push an update to customers, they often fix that push to make it easier. New team members might not know that the team has a process, even if that process is manual, to deliver value to the customers.

When teams have zero hours of overlap, they have trouble creating their shared context, both in project intent and in team capability. The best you can do is to pass tasks back and forth. See *Follow the Sun* on page 44 for this discussion.

Some team members may also consider their personal workspaces and the impact it has on the work of the team.

6.6.3 *Virtual Work Requires a Reasonable Personal Workspace*

Team members need to be able to set the space that allows them to optimize their focus and collaboration with the team.

Many people on distributed teams like to work from home. If so, they need a space in which to work. Both of us prefer that space as a private office, with a door, to keep the family and pets out of the office—unless we invite them in.

The people you hire all over the world may not be able to create that space in their homes.

Just because a person is distributed from the rest of the team does not mean the person works out of his or her home. You might need to provide office space so the person has a place in which to *work*. Consider either paying for or reimbursing an employee's co-working space and a Virtual Private Network.

As a leader, you might need to provide sufficient Infrastructure if you want to have a successful distributed agile team. See *Trap: Underpower the Infrastructure in "Remote" Locations* on page 126.

Beware of adopting what other teams use as a default for your team. Those teams have a different system of work and product context and might need to visualize different work than your team. Don't feel as if you need to use exactly the same tools as another distributed team.

Selecting tools that fit your team's needs requires the team has psychological safety.

6.7 See Your Workspace Traps

Too many distributed teams encounter these traps:

- Someone mandates a specific board for the team
- Underpowered infrastructure
- The team can't experiment enough to create psychological safety

6.7.1 Trap: Someone Mandates a Team's Board

The team's board is the source of its "One Truth": what the team will do shortly, what the work is, where the team has bottlenecks, and what's done. We rarely find that distributed teams can use a prototypical Scrum board as its only information.

Teams need to define and refine their own boards. When someone in power in the organization decides to mandate a board, changing the board becomes a high risk, unsafe decision.

If your organization chose a tool that does not allow a team to easily change its board, consider these options:

- Pick a date when the team will define its own board.
- Make everyone on the team an administrator for the board.
- Ask the team to purposefully experiment with the columns, the WIP limits if they choose specific limits, and everything else on the board.

6.7.2 Trap: Underpower the Infrastructure in "Remote" Locations

One of the reasons we have seen people in "non-headquarters" locations be unable to help the team is that they don't have sufficient infrastructure. Everyone on a team needs the same bandwidth internet access, the same access to tools, the same access to direct one-on-one access to each team member, and the same infrastructure. Both of us continue to see this in supposedly collaborative teams. See more examples in *Lessons Learned from Leading Workshops about Geographically Distributed Agile Teams*, RHI13.

We have seen team members attempt to "share" software licenses because the management was too cheap to buy everyone a license. What happens when a person in one location walks out the door, staying logged into that tool? The sharing person cannot use the tool.

You might need to create infrastructure for the people in the lower-salary areas. If you want these people to contribute to the team, invest in whatever infrastructure you need. We are not advocating bribing local officials to install high speed internet (as one client had to do years ago), but do what you need to do.

Buy people cell phones with whatever speed and data they need. (Not rent, buy. Buy because then people know you won't cut off their capabilities if someone in some Accounts Payable location "forgets" to pay a bill.)

Treat everyone on a team as if they are full-fledged members of the team, with the necessary infrastructure, licenses, and equipment.

6.7.3 Trap: Insufficient Experimentation to Foster Psychological Safety

When teams experiment with their tools—often the cheapest form of experimentation—they build psychological safety for larger experiments (see *Create Psychological Safety in Your Team* on page 88). When other people define the tools, the team has trouble building necessary safety.

Some distributed and dispersed teams are unable to take the responsibility for creating safety in their teams. That's because someone (possibly the Finance people who procure tools) impose a too-narrow boundary of decision-making on the team. Not only does each team member not look for risk, but the team itself has trouble creating the safety necessary for trust and teamwork.

As a leader, consider your possible actions for experimentation:

- Instead of a virtual board, can you suggest the team experiment with a physical board and an always-on camera or a daily image in the team's archival workspace? This might help team members realize everything about their process and product is available for experimentation.
- Ask people to imagine, "If you could change this, what would it look like to fit your needs?"
- How freely can the team experiment with new tools to address current or new problems or provide new capabilities in how they work?

If you have people outside the team setting the team's decision-making boundaries, ask them about their constraints. Often, they have constraints you might not know about. Consider making them allies. Together, you can help your team expand their decision-making boundaries, experiment, and create more safety, *and* help these other people achieve their goals.

6.8 Now Try This

We discussed the strategic intent for a team's tools. A given team's need might change over time. The strategic intent of the team's need for the tool will not change. However, the tools will definitely change over time. That's why we didn't discuss specific tools.

1. How does your team visualize its work? Do they use a physical board or a tool-based board? How well does that board help the team see the flow of its work and where work might be stuck?

2. Does everyone on the team have access to all the tools the team needs to finish the team's work? Can anyone on the team "start up" the tools? For instance, can anyone start a web meeting or does it require one person with a license?

3. Do your team members appear to feel safe in their ability to speak up and suggest alternatives and changes for the team to consider? Do they feel safe in pointing out problems or errors regardless of where they come from?

You have an idea about the tools your team needs. Now, it's time to discuss the team's culture.

CHAPTER 7

Cultivate Your Distributed Team's Agile Culture

Agile approaches change an organization's culture at every level. Agile approaches change what individuals value and discuss—often working more collaboratively than they might otherwise. Agile approaches change how team members treat each other, how people select work, and how they choose to work with each. And, agile approaches often change an organization's culture, from how teams report their progress to how the organization rewards people and teams.

In this chapter, we'll address organizational culture and how you can cultivate your team's *agile* culture.

7.1 Understand Organizational Culture

Edgar Schein (SCH10) defines organizational culture as what people can discuss, how people treat each other, and what the organization rewards. As we discuss your team's and organization's culture, please keep these three areas in mind.

For example, think about the free coffee or juice that is part of a collocated team's culture: that free coffee is a part of how the organization rewards people.

Distributed teams have their own culture. Everyone on the team notices when team members treat people with respect and good intention—and when they don't.

Can people assume the best intention of others, or do they have too few hours of overlap to understand each other? Can people discuss potentially difficult conversations, such as offering and receiving feedback so they understand the intention of the stories, code, tests, and other learning opportunities?

How One Organization Changed Culture as They Grew

One dispersed-by-choice organization, Acme, started with five people, all developers. They knew everything about the organization's choices, they could all write code, and they could test each other's code. One person, the founder, taught the workshops that brought clients into the organization.

The five original people hired more people and grew to 20, 40, and larger. At this writing, the organization is more than 100 people. As the organization grew, the culture changed.

At first, the senior management (the five original people) made all the corporate decisions together. As they grew, they realized they had to share some decision-making. However, once the teams started making decisions, the founders realized the teams didn't have all the information they needed to make good decisions. Neither did the founders. They started to share more information across the organization.

As the organization grew, the management team grew. The founders distributed power into the product, service, and training teams. It was not a smooth transition. As part of their growth, the founders still struggled with how much decision-making power to extend to the new managers.

There is no one "ideal" culture for a distributed or dispersed organization. This company has chosen to make all information open, including financial data and salary data. They try to make it easy for people to discuss—with respect—any issue they have. They have principles, including respect, that help people decide how to work with each other. And, they reward as fairly as they can. They also say that they are a work in progress.

When you think about what people can discuss, ask about the kind of information you and your team has access to. Does your team have access to all the information they require? As in Acme's case, many organizations struggle with who needs which information. The more dispersed a team is, the more information people need. And, the more they need to be able to discuss that information in all their communication channels.

We recommended principles for your distributed team back in *Focus on Principles to Support Your Distributed Agile Teams* on page 23, especially the *Assume Good Intention* on page 29 principle. When people assume good intention and treat each other with respect, they are more likely to create and reinforce a culture of collaboration.

In a collocated organization, we can knock on our manager's door or schedule a meeting if we have a problem to discuss. In a distributed organization, we may need a private backchannel to discuss each person's concerns. Consider creating backchannels between managers and the people they serve, up and down the organization. Those backchannels minimize the risks of insufficient face-to-face time when a topic's urgency does not coincide with overlap in hours.

One-on-ones in a distributed organization are an even more important management tool. (See *Behind Closed Doors: Secrets of Great Management*, BCD05.)

With this context of culture: how people treat each other, what people can discuss (and possibly their access to certain data), what decisions are delegated to teams versus management, and what the organization rewards, are all examples of how you can develop a more agile culture.

For instance, agile teams who select their own approach tend to have a more open culture than teams who are supposed to follow a mandated approach. People decide what they can discuss. They decide how to treat each other. They decide how to reward each other, with the caveat that the managers hold the reward "reins."

That means that an agile approach will change a team's—and the management—culture.

7.2 How Agile Approaches Change a Team's Culture

Regardless of the type of agile approach (iterations or flow), any team moving to an agile approach will change its culture. Here are some common changes we've seen:

From Individual Work *To* Collaborative Work

From Work assigned by someone else *To* Team members select work

From Resource efficiency thinking and metrics *To* Flow efficiency thinking and measures

From Management-planned details *To* Facilitated conversations and decisions

From Gantt charts and other documents as plans *To* Working product and empirical measures to guide further work

From Single-Loop planning *To* Responding to and encouraging change with double-loop planning

From Only understanding product quality at the end of a project *To* Continual learning and improvement of product quality as the team proceeds

When collocated teams move to agile approaches, they have many opportunities to address their team culture.

Distributed teams might need to create opportunities to discuss their new culture. One major challenge we see is when the distributed team moves to team responsibility for the work, not manager-guided individual responsibility.

7.3 Create an Agile Culture with Your Existing Team

Sometimes, a distributed team realizes they can accomplish more valuable work by working together. In this case, they choose to use an agile approach.

For teams with four or more hours of overlap and are new to agile, an iteration-based approach can be a good starting point to understand how to collaborate with rhythm. This not only helps the team focus on the work but also sets expectations of when they will connect online for planning, design reviews, perhaps pair or mob programming sessions, and reviews with stakeholders.

We don't recommend that teams with few to no hours of overlap try to collaborate synchronously as in an iteration-based approach. Too often, that requires time-shifting by one or more people, whichquickly becomes unsustainable.

7.3.1 Create Sustainable Pace

Many agile teams find that sustainable pace supports a team's agile culture and is a key part of how we treat each other. When everyone helps create a sustainable pace, the entire organization benefits.

One team managed their sustainable pace on a case-by-case basis.

One team was distributed across different cities in the same time zone. Occasionally, they would need to deploy their software to customer sites, which could be as far as 12 to 13 time zones away. Deployment took up to six weeks as they installed and tested various scenarios and options with the customer. However, because they were an agile team, they wanted to be responsive to any bugs found at the customer site.

The team members who did not travel to the customer site chose to either shift their work hours or be "on call" for any bug reports. In that same deployment period while the customer installation took place, the "at home" staff worked some very unusual and long hours. However, they quickly resolved any problems and became well known for their attentiveness to the customer. The team was proud of their accomplishments.

When the customer site installation was done, everyone returned home and the management declared a "slow down" period for everyone. Some took a few days off. Others came in but at different hours. Because management gave the teams the space for recovery, they were usually back up and running within a week. Also, everyone knew that another event like this would not happen for several months and they would rotate who would go on site and who would be the "at home" support part of the team.

7.3.2 *Watch for Unsustainable Help*

We more often see servant leaders become stuck in helper roles that are not sustainable.

One notable team had a project manager in Israel with three developers in the Ukraine and one tester in India. In addition, the Product Owner was in Denver, with two more developers in China. Because the hours of overlap were non-existent for the technical people, the project manager and Product Owner collaborated on building the product roadmap, the current backlog, and workshopping the team's stories.

The Product Owner and the project manager were experts in the domain and the code, so they could create the necessary artifacts. These two people worked as servant leaders for the team, but at an unsustainable pace. The team used flow to manage the incoming work and the WIP.

Note that the servant leaders are *not* slave masters, doling out work and pressuring people to work harder and faster. The servant leaders make it possible for the team members to work in an agile way, even if they have few options for collaboration.

Servant leaders are not supposed to be slaves, either. In this situation, the servant leaders were slaves to the organization and the team.

People can work long hours or time-shift their work, certainly for the short term. *Planning* on people time-shifting for the long term is unrealistic.

7.3.3 *Manage Information Persistence as a Team*

Many agile teams find that open access to information is as important as how the team discusses that information. The information is a key part of what we can discuss. When everyone has access to the information, the entire organization benefits.

Each team needs to understand what is critical for their needs for various "information persistence."

What does the team need to know about its product, customers, tools, process, or people? For how long do they need to know that information?

Here are three possible examples of dynamic information: the state of experiments for the product architecture, the state of experiments for the team's process, and notes from current customer conversations.

Teams also have more static information, such as contact information for the people on the team. Teams might also provide tips for troubleshooting known problems with the product.

Note that both of these "static" examples might have an expiration date. Team members may leave the team. The team might resolve these common problems in future versions of the product. Yet, this information does need to persist for a period of time.

Consider this guideline: document the static information and don't be afraid to toss it when we move to a new tool (or change the process or change the people on the team). Decide on a way to keep the dynamic information easily available to the people on the team.

7.4 Build and Maintain Your Team's Agile Culture

When we think about a successful agile culture, we think of teams who can:

- Provide each other with open and honest feedback without fear of reprisal.
- Conduct useful discussions about the product and how the team collaborates.
- Help each other learn about the product and learn to collaborate.
- Work together to finish a limited number of work items frequently instead of keeping everyone in motion.
- Discuss and manage their team's sustainable pace.

However, no agile culture will survive without the support and continuous involvement of organizational leadership. If the leaders do not demonstrate an agile mindset as described in the list above, then it is unlikely that teams will exhibit this behavior. (For more details, see *How Leaders Can Show Their Agile Mindset* on page 227.)

Scale Your Agile Culture As You Grow

Several distributed teams worked on a rapidly growing product. Because they added people, they realized they needed to change how they worked as teams.

The leaders of the group (Product Owner, coach, and engineering manager) met to look at their budget. They needed to plan their hiring and the project portfolio.

The leaders wanted feedback from the teams—did the leaders understand the teams' reality? The leaders presented their information to the teams: roadmap of initiatives, business constraints, and growth goals for this product group. In addition, the leaders created a set of open questions to help the teams think about the entire product experience, from idea to customer experience. The leaders wanted feedback from the teams within a two-week timebox.

In the first couple of days, the teams struggled with these questions. They'd never had this much control over their process. They realized that this was a rare opportunity to shape their work environment and workflow. In the next two days, the teams rearranged themselves from two larger teams to three smaller teams. They did not require the full two-week timebox.

Two teams picked kanban and one picked Scrum. They talked through how they would coordinate as individual teams and developed a backlog of other open questions to work through with the other teams to coordinate across the product group.

When the leadership team met with the teams at the end of the four days, the leaders were happy with the product teams' choices.

As an organization, they decided to have monthly group retrospectives to inspect the current state of coordination across the teams and set new experiments on how to guide the overall direction and development of the product.

Agile teams—collocated or distributed—own their process, their product, and their team organization. When managers serve their teams and encourage this ownership, the team and the organization become more resilient. The team and the organization can change more readily as market conditions and competitors change.

The concept of self-managing teams challenges managers and team members alike. Agile approaches offer the chance to experiment.

We recommend the managers clarify the project charter, keep the team together, and clarify what done means for the project. Then, the managers can trust the teams to deliver.

Learning to trust the teams is even more important for distributed teams. Teams need the ability to solve their own problems. Then, managers become free to tackle the true organizational issues they face.

We find the key to building and maintaining an agile culture is to continue to provide choice to the teams within your organization.

- Choice of where they work.
- Choice of how they work together.
- Choice of tools (within reason and budget constraints).
- Choice of how to collaborate with other teams and groups inside and outside the organization.

Not all teams have the flexibility of these choices. They have decision boundaries for now.

7.5 Understand Your Team's Decision Boundaries

Each team (and each person on each team) needs to understand the boundaries of their decision-making power. For example, teams might decide to define and refine the overall product architecture as a team. A single developer may check with the team to verify their design ideas enhance the overall architecture.

Agile teams need to be able to make—at a minimum—these decisions without external influence:

- What their board visualizes and how the team uses the board.
- The team's working agreements about how the team treats each other and how they collaborate together.
- What "done" means for a story, feature set, release, whatever the product "piece" that the team discusses.

Your team might need other decision-making possibilities.
What might be out of your team's bounds of decision-making?

- If you're not able to use continuous delivery, does the team decide when the team delivers the functionality? Is "done" an intermediate step? Some organizations have a deployment group.

- Tool purchases. If the team doesn't have a budget, the team might depend on other people in the organization to buy the necessary tools. Some members of the team might also have insufficiently powered computers, insufficient communication lines—internet or phone, or insufficient licenses for their work.
- Coaching or consulting or training help. We see many distributed teams who flounder because they are unable to allocate money for coaching, consulting, or training. Their management doesn't realize how valuable a little extra help might be.
- Language/compiler choice, development frameworks, or version control systems. Many developers and testers like to use the "next best thing" especially when it comes to languages. Organizations might not want to invest in the next best computer language. That decision might be out of the team's purview.
- Choice of new teammates. We both advocate having the hiring manager facilitate the hiring process and ask team members to participate in the interview and decision-making process for who to add to a team.
- Choice of removing team members from a team. Because most HR functions require management involvement, it might not be easy for a team to remove a team member.

Use the principles of collaboration, experimentation, and the agile principles to expand those boundaries where reasonable. Help your teams discover how to deliver the best results.

7.6 Enable a Collaborative Distributed Culture with Helper Roles

Many people, including managers and team members, believe that tools can make distributed teams successful. Not true. Many people believe agile approaches can make distributed teams successful. Maybe.

One of the major elements of a successful distributed agile team is the ability of the team to discuss anything the team needs, and to treat each other in a way that promotes collaboration.

We like to think about team collaboration this way:

- Do the people treat each other in a way that enhances the team's collaboration?
- Does every person on the team participate in team-based experiments?
- Does everyone on the team have equal opportunity to contribute to team conversations?

These are just three questions that might help you think about your team's culture and necessary environment. You might need more questions.

Your team might need to work with other teams, sponsors, or customers. To maintain your agile culture, consider if you need any helper roles, such as copilot, ambassador, or buddy. We recommend people rotate through these roles, as well as rotating through the role of an integration buddy when hiring.

7.6.1 Consider a Copilot or Proxy Facilitator

The more locations your team has, the more your team might need a copilot or a proxy facilitator to support the team. Copilots bridge the distance and time between team members.

In planning, a copilot may assist a coach or Product Owner by monitoring the backchannel for questions or comments while other conversations occur on the audio channels. A copilot might also assist with technical issues with tool access or meeting software before or during a planning meeting (or any meeting for that matter).

However, copilots can be useful outside of meetings as well.

A Team Member Serves as Copilot

A Scrum Master, Jim, was in one city while the rest of the team was in another city within the same time zone. Jim struggled to teach and help the team with their initial agile adoption and transformation.

After a few weeks, Jim discovered that one of the senior developers, Jane, who had the respect of the team, also was a good listener and paid attention to the team's morale. Jim decided to then sync up daily via a phone call with Jane to hear what she observed in the daily standups and meetings. How did the team react to new stories? Did they seem puzzled? Confident?

Jane read the body language of the other team members and helped facilitate some of the team conversations. Eventually, Jim helped teach agile techniques and facilitation to Jane so she could assist the team more.

The copilot can be any member of the team, another coach, Scrum Master, or agile project manager, who can help create additional hours of overlap to facilitate the team and their processes.

Copilots help in situations where people are reluctant to explain they don't know something, or cannot engage with the team. A co-facilitator helps everyone understand the progress—or lack thereof—a team can make.

7.6.2 Use an Ambassador

An ambassador can also help connect the different team members.

The ambassador is any team member who visits other parts of the team. A team member from one location visits part of the team in another location for a week or more. While there, the team member works and socializes with that part of the team. The "ambassador" then returns to their home location.

The visit creates a personal connection between the people in both locations. They help others understand the "context" of the other location. For instance, Ian in the UK may be occasionally late to certain meetings because he has several children to pick up from multiple schools. The ambassador helps share this understanding with other parts of the team.

Other teams have the opportunity to use digital nomads as ambassadors. Some people like being a digital nomad, taking extended road trips. These people visit with different team members, working as it makes sense. Sometimes, they pair, in their homes (with permission), in coffee shops, or in coworking spaces. Digital nomads can become part of the social glue that connects the organization.

The ambassador does not need extensive facilitation skills. Their *personal* connections across the team will serve them well in facilitating conversations.

7.6.3 *Use a Buddy*

Instead of a copilot, you might choose to use a "buddy system" among your team members to help them stay connected.

In meetings of satellite or cluster teams, you might have each person in the main meeting room pair with a remote participant as a buddy. The buddy's job is to listen to the remote buddy and bring their issues to the meeting in the main meeting room.

Pair development is another form of the buddy system. When two team members learn together and co-facilitate each other's work, they are more effective than either person alone. In addition, they tend to improve the overall quality of the work.

Consider if your team wants to use a buddy system when hiring new people. See *Use a Buddy System to Integrate New People* on page 212 for more details.

7.7 See Your Team's Agile Culture Traps

We've seen several culture traps in distributed agile teams:

- One or two people bridge time zones for the entire team.
- One cluster dominates the discussions.
- Leaders assign work to specific people.
- The team assigns work to specific people.
- [a] Some decisions surprise some team members.

While we provide some suggestions below to avoid these traps, consider how your team can decide how they will work together; what issues they will discuss and how they will discuss them; and how they might recognize each other for the maximum benefit to the team.

7.7.1 Trap: One or Two People Bridge the Time Zones for Everyone

We saw how the servant leaders back in *Watch for Unsustainable Help* on page 134 took the brunt of the few hours of overlap.

We do not recommend teams work like this. In fact, we would say anytime any team member feels they need to bridge the gap of insufficient hours of overlap, this is a culture trap.

Please read *Follow the Sun* on page 44 for information about how you could make insufficient hours of overlap work. You always have the option of using a more traditional project approach. Also, see *When Agile Approaches Are Not Right For You* on page 17.

7.7.2 Trap: One Cluster Dominates

Sometimes one cluster dominates conversations in synchronous team meetings. The other cluster(s) on video or audio appear to be "just" listening. The team members might be active on the backchannel, and it's difficult to tell if they are engaged. The team does not appear to have a collaborative and agile environment. We've seen this occur most often when the PO or a leader is part of one cluster.

Consider a local copilot for each cluster who can encourage participation or, at least, give hints to the primary facilitator on how to engage the other clusters in collaboration. You might consider asking people to *Use a Buddy* on page 142 across the clusters or *Use an Ambassador* on page 141 to reach other teams.

Additionally, consider speaking to the other cluster(s) on how to create a work plan for the rest of the team. Here are possibilities:

- Share the vision for the initiative or feature.
- Create small stories, with no tasks.
- Ask the others to explain how they prefer to generate their work plans.

7.7.3 Trap: Leaders Assign Work to Individuals

We've seen Product Owners (or even managers) who use a tool to "contract" the work to individuals, reinforcing each person's functional expertise. We see more of this sort of assignment to individuals when teams start with a tool for their board and backlog. The tool creates or reinforces a mental model of resource efficiency instead of flow efficiency.

Because distributed teams tend to collaborate through a tool, they tend to encounter this trap more often.

We are not opposed to tools. We are opposed to assumptions and tool use that reinforce a resource-efficiency culture instead of a flow-efficiency culture. We are also opposed to trying to maintain a single person's expertise because this is just another form of resource efficiency.

Encourage your team members to collaborate on the work, so that every team member learns about all the other areas of the product. See *Collaborate on the Work to Move it to Done* on page 185 for more specifics.

7.7.4 Trap: The Team Assigns Work to Individuals

Some teams reinforce each person's expertise with a rule that says only people with specific knowledge can develop or review the code or tests in a specific area. The experts become a bottleneck for that area of the codebase and for the entire team. One way to resolve this is to pair or mob on that part of the code to educate other members of the team. If the team is not comfortable with pairing or mobbing, then it is worth exploring ways to educate other team members on the code as part of their day-to-day work.

Other teams like to use code review to manage knowledge gaps and knowledge transfer about many areas of the code. This enables many people to review much of the codebase.

If a team has insufficient hours of overlap, the team will reinforce the experts' expertise. See *Measure Costs of Delay in Distributed Teams* on page 228 for more discussion.

7.7.5 Trap: Team Decisions Surprise Some Team Members

If you have a satellite team member who seems to be surprised by team decisions, you may not have the optimal environment for the entire team to be successful.

We see this often when one team member has very few or no hours of overlap. In that case, we recommend that both locations consider themselves "remote." We suggest the organization build a cross-functional team in each location. We have yet to see one lone person, in any project approach, with few to no hours of overlap succeed in meeting the team's goals.

Cluster teams might devolve to collocated silos of people if the team doesn't tend to its culture, instead of a cohesive team. Clusters might need a different supporting environment than the satellite teams.

7.8 **Now Try This**

Every distributed agile team creates its own culture, for the benefit of the entire team. Consider the following ideas:

1. Assess your team's culture based on the *From-To* list in *How Agile Approaches Change a Team's Culture* on page 132. Are your team's behaviors more on the "From" side or more on the "To" side? What can you do to help shift everyone's behaviors to those more on the "To" side? What behaviors does your organization reward now? Do you need to start those difficult conversations about what the organization rewards?

2. What decision boundaries does your team have? Team members need to know what they can and cannot discuss for now. Which boundaries are movable?

3. What is the supporting environment your team has available and what tools are available to the team. Does everyone have equal access? Does everyone on the team have the ability to "reconfigure" the tool? Does the team have access to other people who can help support their work like co-facilitators or copilots? If not, do team members have training where they can take on these roles?

Next, we'll discuss how to build respect among the team members with working agreements.

CHAPTER 8

Build Respect with Working Agreements

Too often, distributed teams (agile or not) start by dividing up the work. In these less-than-optimal distributed teams, each person works away quietly until "their part" is done.

The team brings the pieces together, often when the testers start their work. Surprises abound! Too often, what the team created is not what's in the Product Owner's head, and does not meet the acceptance criteria.

That's because the team members didn't collaborate. They worked "at" each other, not "with" each other.

The more the team continues this pattern of surprise and insufficient collaboration, the less respect each team member has for the others. Add the friction of insufficient hours of overlap, and the team heads towards implosion.

Agile team members work best when they work "with" each other and not "at" each other. Working with each other in a more collaborative fashion helps build and maintain respect for other team members, especially with fewer hours of overlap.

More successful distributed teams start by building respect for each other as people. The team members realize that each person has their background, skills, and opinions. When team members recognize each other as humans, they can leverage each other's strengths and build safety in their team. They can see each other's opinions. Even if they may not agree on the approach, they will seek out diverse opinions on the best approach.

With respect in place, the team can navigate these options and decide on what will be the best approach for the team and the product.

Finally, the team needs to respect each other's context as they collaborate. Understanding where work hours overlap or do not and understanding personal versus work commitments can help the team navigate the day-to-day collaboration to build the best solutions for the work at hand.

8.1 Lack of Empathy Can Prevent a Team from Norming

One team had five five people over three time zones across the U.S. The team members had never worked together before and had never worked on a distributed team before. They were new to the product. Each team member had been selected by their manager as the "best" for their particular technical expertise.

None of the team members was particularly empathetic. This might not have been a disaster, but the team members did not meet each other until after the project started.

As the team started to work, each member focused on their own expertise: platform, middleware, app layer development, UI, and system testing.

The project never had a kickoff meeting. There was no project charter. The team members had instructions from their manager to "do the agile thing" and finish the product.

The team finished nothing in the first month. Luckily, each team member complained to their relevant managers, and the managers recruited a person they called an "agile project manager."

This project manager had one-on-ones with each person and realized the team wasn't acting as a team at all. The project manager insisted on bringing all the people together in neutral territory, a hotel meeting room. Neutral territory means no person has an affiliation to the physical location—the meeting space is *not* part of any team member's normal workspace. This allowed the team members to think outside their normal workspace and habits.

In the meeting room, they created a project charter and a product roadmap first. They then workshopped their initial stories. (The lack of a Product Owner is a different part of this story.)

In that week together, the team learned how to work together as a team. During that week, the project manager had to say several times, "How would you feel if someone said that to you?"

Each person was very smart and was accustomed to being the technical lead on their teams. They had to learn how to work together, listen to each other, and feel what the other people felt.

This team eventually finished the project and released the product. It was painful even after they spent a week together, because none of the people had been added to the team based on their empathy skills.

Most of the people we meet in technical fields discover the technical field because they like the technical problem-solving, a problem-oriented perspective. Few technical people we meet are people-oriented *and* problem-oriented. Yet, distributed teams require more diplomacy and empathy than collocated teams do.

8.2 Distributed Team Members Require Empathy

Empathy is the ability to understand and share the feelings of another. Empathy is not sympathy, where people feel pity or sorrow for the other person.

We find that many people can develop empathy if they ask this question:

What would have to be true for this person to act or react in this way?

A new Scrum Master moved to the "A Team," where a well-respected architect, Arthur, appeared to do very little work. The team respected Arthur's reputation and let his lack of work slide—for a while. After several months, other team members started to resent his lack of work.

In a retrospective, the Scrum Master asked, "Is there one item we haven't discussed?" At first, the team was silent. Then, Arthur spoke. "My wife is fighting a terminal illness. I've been spending time with her." He apologized to the team and asked for help. How could he avoid slowing the team down?

The team's demeanor changed immediately. Instead of animosity, each team member moved to complete support. Team members asked, "How can we help you?" and "What do you need?"

After further discussion, the team decided to add to their working agreement what they called the "three strikes rule" (some were baseball fans). If anyone on the team had the same update to the team after three daily standups, anyone on the team could offer help with no questions asked as to why the person needed help. This took tremendous pressure off of Arthur and further increased the empathy and respect among the team members.

Note that Arthur asked for help. When team members ask for help, they show that they are strong, not weak or vulnerable. When team members feel that they can ask for help, they reinforce the *team's* responsibility for the work, not the individual's.

8.3 Asking for Help Can Build Respect

Too often agile team members—because of organizational culture—don't ask for enough help. People don't realize that asking for help is a gift to the other members of the team.

Your team might benefit from a board that shows Asking for Help or Offering Help.

Ask/Offer Help

Person	Day 1	Day 2	Day 3	Day 4	Day 5	Day 6	Day 7	Day 8	Day 9	Day 10
Sylvie		A			A, 0	A	A	0	0, A	A, 0
Jacques	A, 0		A		0	A	A	0	0, A	A, 0
Stefan	0				0	A	A	0	0, A	A, 0
Cindy	A	0			A	0, A	A	0	0, A	A, 0
Tom		0	A		A	0, A	A	0	0, A	A, 0
Mike		0	0		A	0, A	A	0	0, A	A, 0

Figure 8.1: Possible Collaboration Board

During a team retrospective, one team examined the problem of "Why are we completing stories so late, if the stories are so small?" They realized each of them had information the others needed. They decided to track their collaboration using this board. They started with a daily tick-mark in the form of an "A" for ask and "O" for offer.

They decided to review the board at their standup each day as a way to remind themselves to ask for or offer help. As they reviewed this board, they realized they asked for and offered help intermittently.

On Day 4, they didn't ask or offer at all. The board prompted them to ask themselves why. They realized they had several problems:

- They had too much WIP. Everyone was on a separate story for multiple reasons.
- The "too much WIP" problem helped them realize why everyone was too busy to respond quickly enough to a chat request.
- On Day 7, they realized all the work they'd done independently now needed other people.

They used this board to address their team's WIP and to prompt themselves to ask for help more often.

Some teams decide not to wait for three days, as Arthur's team did. Some teams have an agreement of "If you're stuck for 15 minutes, ask for help."

8.4 Facilitate Decisions About Respectful Teamwork

Agile teams have several tools to create respect and empathy in their teamwork:

- Understand each other's personal preferences for work and time.
- Understand the team's values.
- Create team-based working agreements.

Let's start with personal work preferences about how people work and when they work.

8.4.1 Discover Personal Work Preferences with the Compass Activity

Each of us prefers to work in certain ways and at certain times of the day. You can discover these preferences through several different approaches. You might be familiar with various personality assessments. We recommend you start with a lightweight assessment, such as the *Compass Activity for Distributed Teams* on page 257, which might be just enough information for you and your team.

In the Compass Activity, team members self-identify as one of the following:

- North: Acting—Always in motion. Likes to try or plunge in.
- East: Speculating—Like to examine the big picture and options before acting.
- South: Caring—Likes to know that everyone's voice has been heard before acting.

- West: Detail-focused—Likes to know what, who, when, where, and why before acting.

Preferences are not destiny. They are our *preferred* work habits only. Given new information, we might even change our preferences.

8.4.2 See When Each Person Works

Some people like to work early in the day. Some prefer to work later in the day. Some people work Sunday through Thursday and some work Monday through Friday. When people explain *when* they work with respect to the rest of the team's time zones, it's easier to see the hours of overlap for collaboration.

One way to see this is with a time zone bubble chart, as in *I'm Working While They're Sleeping: Time Zone Separation Challenges and Solutions*, ERR11.

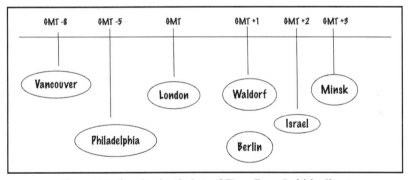

Figure 8.2: One Project's Actual Time Zone Bubble Chart

The bubble chart shows each person's native time zone. It does not show when people actually start and end work during the day and which days they work.

People can and do take time for themselves during the day. Distributed work can lend itself to more flexible personal schedules. Your team might choose to improve this chart by adding other time preferences, such as lunch or workout time that people might take in the middle of the team's hours of overlap.

Jane, a developer on a distributed team, takes workout time during the day. Part of her workout is to "punish" herself, in her own words, with squash. On Tuesdays and Thursdays, she takes two hours in the middle of her workday to play squash with a friend at her gym.

TUESDAY, THURSDAY													
City / Person	local hours of the day / hours worked by team member												
Raleigh	5	6	7	8	9	10	11	12	13	14	15	16	17
Sarah				1	1	1	1		1	1	1	1	
Mary					1	1	1	1		1	1	1	1
Boston	5	6	7	8	9	10	11	12	13	14	15	16	17
Jane				1	1	1	1			1	1	1	1
Mike			1	1	1	1	0.5	0.5	1	1	1		
London	10	11	12	13	14	15	16	17	18	19	20	21	22
Ian	1	1	1		1	1	1	1	1				
Berlin	11	12	13	14	15	16	17	18	19	20	21	22	23
Gerrit			1	1	1	1	1		1	1	1		
Overlap Probable	17%	17%	50%	67%	100%	100%	92%	42%	67%	83%	83%	50%	33%

Figure 8.3: Jane's Team—Tuesday and Thursday Work Hours of Overlap

When she leaves, she says, "I'm going offline now, to punish myself again." Her team knows when to expect her to return and that she will be pink-faced and happy, ready to collaborate again. Jane's team uses the enhanced time zone chart to see everyone's availability.

MONDAY, WEDNESDAY, FRIDAY													
City / Person	local hours of the day / hours worked by team member												
Raleigh	5	6	7	8	9	10	11	12	13	14	15	16	17
Sarah				1	1	1	1		1	1	1	1	
Mary					1	1	1	1		1	1	1	1
Boston	5	6	7	8	9	10	11	12	13	14	15	16	17
Jane					1	1	1		1	1	1	1	1
Mike			1	1	1	1	0.5	0.5	1	1	1		
London	10	11	12	13	14	15	16	17	18	19	20	21	22
Ian	1	1	1		1	1	1	1	1				
Berlin	11	12	13	14	15	16	17	18	19	20	21	22	23
Gerrit			1	1	1	1	1		1	1	1		
Overlap Probable	17%	17%	50%	50%	100%	100%	92%	42%	83%	83%	83%	50%	33%

**Figure 8.4: Jane's Team—Monday, Wednesday, and Friday
Work Hours of Overlap**

Team members, while they might have a full day of overlap, do not—and should not—be tied to their desks, computers, and team members the entire day. The team needs to be able to see when people are available to collaborate, just as they would if they were a part of a collocated team.

Some team members might decide to offer a small time shift, to create more hours of overlap with their team. We recommend that team members choose this possibility for themselves. If someone with hierarchical power or authority mandates this option, people resent it.

8.5 Identify the Team's Values

In order to create effective working agreements, consider the team's values.

Dhaval Panchal[1] has a terrific team exercise to articulate the team's values:

1. Ask everyone to meet for about 30 minutes. (This assumes the team has at least 30 minutes of hours of overlap.)
2. Provide everyone with index cards and a large, dark marker. The team members will need to see each other's cards after writing them. (If the team is not able to meet in person, consider using an online app or a shared spreadsheet or drawing.)
3. Ask each person to fill in this sentence: "I don't like it when someone/people . . ." Each person might write down anywhere from two to five of these sentences.
4. Divide the team into pairs.
5. In pairs, select one card. Work with your partner to write down a statement that counters the negative statement. For example, if you wrote, "I don't like it when some people tell me what to do," you might write a statement that says, "I like it when people discuss our technical approach as a team." Continue until each pair addresses all its cards.
6. Ask the pairs to read each "I like it" card out loud. As the team members read the cards, capture what they say on a flip chart

---

[1] http://www.dhavalpanchal.com/sharing-values-a-team-building-exercise

to post in the team area. (Note: the facilitator does not read the "I like it" statements. The team members read them.)

If this value discovery exercise does not work for your team, consider researching others. You can adapt many of the exercises written for collocated teams to distributed teams.

8.6 Create Working Agreements

Working agreements create a team's culture. Teams decide how the team members work together as part of their working agreements.

The team might address these kinds of issues in their working agreements:

- What "done" means. (For more information about defining all the possible "dones," see *Create Your Successful Agile Project*, ROT17.)
- Ground rules of meetings, for example when people talk, listen, and mute themselves in various size meetings. (See *Work Together Anywhere*, SUN18 and *Manage It!*, ROT07 for tips on meetings.)
- Define hours of overlap for collaborative work. See *Establish Acceptable Hours of Overlap* on page 24.
- What sustainable pace means for this team. See *Create Sustainable Pace* on page 133.
- Norms about personal focus or personal time. See *Create Resilience with a Holistic Culture* on page 33.
- How the team communicates, especially with video, chat, email, and more. For example, "Don't use one-on-one chat unless it's personal or urgent."
- Which tools to use and how. See *Create Your Collaborative Team Workspace* on page 111.

There are two parts to working agreements: how each person prefers to work and the team's agreements for the team's approach to work.

8.7 **Blend Personal and Team Working Agreements**

We suggested that team members learn about each other in the *Compass Activity for Distributed Teams* on page 257. That's because team members who work on teams with fewer hours of overlap will tend to more solo work.

With any luck, your team has enough hours of overlap to collaborate. If so, the team needs to blend their personal working preferences with the rest of the team members.

We have seen distributed teams address their working agreements in a facilitated meeting where everyone is in their own separate office location. That's okay, and can work.

Instead, we strongly recommend the distributed team meet in person for a minimum of one week. See *Non-Collocated Teams Deserve Face-to-Face Time* on page 12. During this time, we recommend the team use a short daily retrospective to see if their process works for them. The team can then understand what to do.

Not all teams can meet and work together for even one week. If not, consider facilitating the team's value statements, and some form of working agreement activity. We recommend timeboxing these meetings to not more than an hour each. You might consider a 30-minute timebox for each meeting, to maintain the team's focus.

If the working agreement meeting has too much content for one 30-minute timebox, create several shorter meetings to address each part. It's fine to have a 10-minute meeting about core hours and a 20-minute meeting about what done means, and a 5-minute meeting about meeting times, and so on. Consider having breaks between each of these meetings. Or, consider addressing a different part of the working agreements every day for a week.

For a distributed team to accomplish this while distributed, the business partners and management need to understand the teams will "slow" their work while the team creates and reflects on their working agreements. Consider this time investment as "regular maintenance" for the team just like you may go for a regular doctor's visit. When you make time for these adjustments, the team learns how to proceed better and faster.

8.8 Define the Project Charter

In addition to working agreements, each project needs a charter, with at minimum the project objective and release criteria. (See *Create Your Successful Agile Project*, ROT17, for more detail about how to write a project charter.)

We recommend someone help facilitate the team when they write the project charter. That someone could be the Product Owner, an agile project manager, Scrum Master, or coach. While we recommend the team create its own project charter, teams might have questions that the facilitator can help resolve.

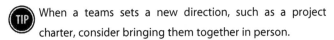 When a teams sets a new direction, such as a project charter, consider bringing them together in person.

Not every agile team has a "product" as its primary output. Some agile teams provide customer, operations, or production support. Those teams might not have a project objective or release criteria. Instead, those teams tend to have service level agreements. See *Around the Clock* on page 42 for more information.

8.9 Consider These Tactics to Build Teamwork

Teamwork itself is not the goal. The goal for agile teams—collocated or not—is to create the best product for the customer in the least amount of time. Teams do that with flow efficiency, learning how to be a team, collaborating with each other.

Non-collocated teams can collaborate on the work, when they live the eight principles. Consider these tactics to build collaboration and throughput:

1. Bring people together for a few days to a week to learn how to work together. See *Non-Collocated Teams Deserve Face-to-Face Time* on page 12.

2. Teach team members and leaders how to offer and receive feedback as in *Behind Closed Doors: Secrets of Great Management*, BCD05. The sooner team members can provide valuable feedback to each other, the less likely that team conflict will fester.

3. Virtual Lean Coffee: A lean coffee meeting generates a ranked agenda with the people present. Then, the people use short timeboxes to focus their concerns and ideas on one topic at a time. They decide to continue or change at the end of the timebox. See[2] for the template of how to create a collocated lean coffee meeting. Mark developed a distributed lean coffee template.[2] Lean coffees can help team members discover and create bonds within teams, across teams, and start to facilitate Communities of Practice.

Also, keep in mind that each of these tactics might be used in different ways.

8.10 Build Respect Across the Organization

In addition to building respect within your team, consider what you might need to do to build respect across the organization.

We explained that sometimes the organization optimizes for individual-as-silo in *Trap: Leaders Assign Work to Individuals* on page 144. When people don't collaborate on the product as a whole, Conway's Law kicks in: the product has multiple ways to pass and receive data; the user experience isn't consistent; and the overall architecture isn't coherent. The product development teams complain about needing the software equivalent of duct tape to make the product work. They continually request to re-architect the product.

When we optimize for respect, we develop habits within each individual to ask for help from anywhere in the organization and

---

[2] http://leancoffee.org

[3] http://markkilby.com/virtual-lean-coffee

help develop a more holistic culture (*Create Resilience with a Holistic Culture* on page 33).

Sometimes, the optimization is not for the *team*, but for a specific location, often a cluster. In this situation, the individual silos are replaced by location-based silos of collaboration. This is a slight improvement, but can still lead to unnecessary complexity in the products. The user interfaces for various products or parts of products are just a little different from each other. There are too many similar-yet-different access points for the database(s). The middleware has parallel options for the same results. Conway's Law still exists.

Instead of optimizing for a person or a cluster location, optimize for a *product* or *feature* team. We have seen feature teams in many locations succeed—to deliver outcomes of entire features.

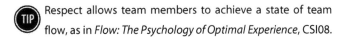 Respect allows team members to achieve a state of team flow, as in *Flow: The Psychology of Optimal Experience*, CSI08.

Optimizing for respect helps the team(s) optimize for the entire product, not just "their parts." The teams are more likely to create products with consistent user interfaces, provide a consistent maintenance experience, and allow various types of users to quickly accomplish a wide variety of tasks.

As a leader, consider your options for creating respect within and beyond your team. You might need to encourage respectful and trustful discussions. Help people catch themselves, both when they don't create respect and when they do.

8.10.1 *Explore Tactics for Encouraging Respect Across the Organization*

As a leader, look for opportunities to develop and reinforce respect and empathy across the organization:

1. Encourage teams and team members to create and use dedicated team backchannels to discuss issues important to their work. Encourage the entire team to discuss common issues across

the organization. See *Practice Pervasive Communication at All Levels* on page 28.

2. Encourage and plan events (e.g., hackathons) that allow individuals across the organization to explore new skills with the understanding that they should share what they learned. The more people try alternatives and share their results, the more the rest of the organization learns about these people and respects them. Their experiments don't have to work. When they share their learning and/or ask for help, they show respect and empathy and they earn the same.

3. Work with managers to explain the costs of working in resource efficiency. (See *Think in Flow Efficiency* on page 10.) The costs are not only in the delays of work. The people suffer the intangible costs of lack of respect, which leads to technical incoherence in the product.

4. Offer reinforcing feedback when you see people working with respect and empathy.

5. In a retrospective, call attention to the *team's* product and relationship progress. You might help the team notice their cycle time and their collaboration board.

You may see other opportunities to capitalize on building respect and empathy in your organization.

8.10.2 *Re-Evaluate the Value of a Bully*

As a leader in your team, consider how people act on behalf of and toward the team. People often have good intentions. They want to use their expertise to do a great job for the team. They may have contributed to multiple teams and they may feel highly invested in the company's success.

Their expertise creates the illusion that they are "indispensable" experts. Their technical contributions do help the team. However, their interpersonal contributions can destroy respect.

One program had this bullying problem across multiple teams.

BuildIt had distributed cross-functional teams across Europe and the U.S. Each team was cross-functional in its own location, and they collaborated as a program to deliver the product.

BuildIt spun off its flagship product into two different products: Product Lite and Product Pro. The idea was they could use most of the codebase for both products and turn off some of the features in Lite that they would use in Pro.

Management asked the most experienced teams to work on Pro. The least experienced teams worked on Lite. When teams split on experience levels, the people on the lesser-experienced teams don't have enough access to the people with the answers. That creates Costs of Delay (see *Measure Costs of Delay in Distributed Teams* on page 228). That Cost of Delay increases cycle time for the lesser-experienced teams to make progress on their part of the project, program, or product.

 Maintain a mixture of experience when you split teams.

Unsurprisingly, the overall product architecture drifted apart in the very first month. One of the Pro developers used the dedicated team backchannel to respond to questions from the Lite developers in this way, "Any idiot would know that, and here is the obvious answer ..." and "Only a raging lunatic would ask that question, and here is the obvious answer ..."

The program manager noticed this behavior and first had a one-on-one conversation with the Pro developer. That changed his behavior for about 24 hours. She had a second conversation, with the same effect.

She spoke with his team. While they agreed with his frustrations—management had requested a too-early date for initial release—they disagreed with the way he expressed his frustrations.

The program manager asked for an all-program meeting of the seven teams. The agenda for that meeting was to define the agreements for architectural and design decisions; how people would discuss problems in the dedicated product backchannel; and what a reasonable release date might be.

Management was concerned about the expenses for an all-program meeting. The program manager showed them the cost of hiring more people to replace the two teams she thought might leave. Management agreed to the meeting.

At the meeting, the Pro developer asked to be the first agenda item. He apologized in public and specifically apologized to the several developers on the Lite teams he had offended. That apology allowed the team members to recreate an environment of civility and respect.

The teams still had trouble with the problems they needed to solve. However, without the bullying, they were able to do so.

This story had a happy ending, with the Pro developer rebuilding respect with his peers. Too often, we don't see a happy ending.

When people bully others, they create resentment. This program manager saw the two teams' resentment and acted. Bullies might not create a cost you can easily see on a value stream map. They create a cost that seeps into people's morale over time. Bullies create an environment that violates the principles of *Assume Good Intention* on page 29 and *Create Resilience with a Holistic Culture* on page 33.

Encourage everyone to be aware of potential bullying and offer feedback. Not every bully realizes the effect of his or her behavior.

8.11 Work with Humans Requires Empathy

Neither of us—along with many of our colleagues—entered technical fields because we were people-people. We started in technology because we loved to solve problems with technology. We were surprised by how much of our work is not about technology—it is about interaction and collaboration with other people.

We have seen many people who excel at solo problem-solving. They may not excel in collaboration for any number of reasons.

They can be good in a team if they can find the "right" team. Often, that team excuses not-so-empathetic behavior because the team member is willing to coach and mentor others in their expertise.

We have worked with people who miss emotional cues, at least, the first time those cues arise. Sometimes, they never see those cues. That means their empathy is lower than others.

Some people prefer distributed work because they can protect their work environments for effective solo work. They could control every aspect of their solo work: how they took in requirements, how they designed and tested, and how they released. They could control their interactions with others.

These people have controlled their own environments, possibly for years. Now, with an agile approach, we ask them to change everything about how they work. They might not feel comfortable in the team, with their work.

 Watch for people new to a distributed agile team whose satisfaction degrades over time.

People new to a distributed agile environment might need more empathy. And, they might need more than just empathy.

Consider offering private and public conversations: one-on-ones, an "ask me anything" session, or a lean coffee where anyone in a small group can raise any topic. Consider articulating how each person helps the entire team. You might need to do this in public, with appreciations, or privately, in a one-on-one.

Help the people who had been successful as solo contributors learn how to fully participate as a team contributor.

We know that different people respond to large and small changes differently. If you, your team, and your management have tried to accommodate a given person's need for managing change, and it's not working, consider these options:

The rest of the team may find it helpful—for now—to ask this person to work solo on special projects. This might not be a long-term solution.

- Make sure you have a one-on-one conversation (with video or in person) with this person to understand his or her concerns. Start with curiosity so you can hear the other person.
- Emphasize what's valuable about the change for this person. If this person continues to struggle working in the changing team environment, you may explore placing them elsewhere in the organization. The rest of the team may find it helpful—for now—to ask this person to work solo on special projects. This might not be a long-term solution.
- If the changes provide no benefit to this person, maybe it's time to help this person find a new role outside your organization.
- If the person's responses seem "off" or too extreme for the situation, consider bringing in HR, or suggesting the Employee Assistance Program (or the relevant program in your country providing counseling services or access to those services).

In some ways, a distributed agile team requires even more from each person. In exchange for the freedom to work "anywhere" (within some bounds), team members need to collaborate and empathize with each other.

While distributed agile teams can accommodate individual preferences, the team members optimize for the entire team.

8.12 See Your Respect Traps

We've seen too many respect traps take teams out of their flow:

- Create an elite team and never bring them together in person to build respect and learn how to work together.
- Assume offshore team members don't need respect.
- Frequent or unexpected team membership changes.
- Allow bullies to win every argument about the project and the product.
- Assume vendors are not part of your project or program.

8.12.1 *Trap: Create an Elite Team and Never Bring Them Together*

Managers think they can assemble "expert" distributed teams when they collect the people as components: the code architect (or senior developer), the test architect (or senior tester), and several other experts in their areas.

One engineering organization needed a high-innovation, initial product release completed in six months. The managers chose their team: the architect in China, the middleware expert in France, the UI expert in Denver, the database modeler in San Francisco, and two senior testers in Hyderabad.

Each of these people had extensive experience on previous versions of the product. They knew of each other, and had respect for each others' accomplishments.

However, these people had never worked together in person before.

They tried to work as a team. They had insufficient hours of overlap to all meet when everyone was awake. They tried time-shifting for a couple of weeks, but each of the people had family obligations that prevented them from time-shifting for long enough to really form as a team.

They asked for a budget to travel somewhere in the world to all work for a couple of weeks together. They estimated their travel and hotel costs at about USD $40,000. Management said no.

The team struggled for six months and never quite completed their first deliverable. They tried to use all the ideas in Follow the Sun. The cost of delay for their product was at least USD $500,000. Yes, the company lost that much money because management did not spend a tenth of that on bringing the team together.

Consider these questions before you decide travel is "too expensive" for a team:

- Why do you need the experts? Is it their expertise? Might others with similar expertise but more hours of overlap help create a team with more throughput?
- Is it worth creating a team with more hours of overlap rather than experts in each area?
- What is the value of the product this team will create? Can you calculate the Cost of Delay of a month of this team's time?

You might have better results with these options:

- Ask the experts to distribute their expertise across the organization. That might mean pairing or other team-based work with others who have more hours of overlap.
- Create teams with sufficient hours of overlap, especially for high-innovation product development.
- Bring people together for a week or two to help them learn how to work together. See *Non-Collocated Teams Deserve Face-to-Face Time* on page 12. *Consider Handoffs* on page 59 when they are apart.

Whatever you do, don't let teams thrash for months when they need to learn how to work together.

8.12.2 *Trap: Assume Offshore Team Members Don't Need Respect*

We've seen many teams who have "offshore" employees. Some teams have offshore employees and some have contractors. Regardless of who or where those teams are, all team members need to build respect with each other.

In agile approaches, the team is the working unit, not a single person. That means the team needs to have consistent team membership and ways to build trust and respect with each other.

When people persist in thinking about where people are, instead of their team membership, they create an atmosphere of disrespect. This is another case of resource efficiency thinking that can destroy any possibility of respect on a distributed team. Instead, *Think in Flow Efficiency* on page 10.

8.12.3 Trap: Frequent or Unexpected Team Membership Changes

Some organizations create dispersed teams with outsourced members. Too often, those people are testers.

As they learn the product and how to do their job, they become more valuable to the outsourcer. These people move on to other jobs, often on other products. Sometimes they move to other outsourcing companies.

The product team experiences rapid turnover of necessary people. The team loses its tester (or whomever) on a regular basis. The existing team continually retrains the new team member.

That continual retraining exasperates the existing team members. Over time, they lose respect for the new team members, or anyone from the outsourcer.

Respect takes time to build and is fast to destroy. When a part of a team feels as if they waste their time continually retraining other members of the team, they stop treating the new people with respect.

Rethink your organization's contracts with outsourcers, to integrate *Think in Flow Efficiency* on page 10 into the contracts. In fact, you might find that outsourcing is not what you need. You might need feature teams all over the world, integrated as teams into your organization.

8.12.4 Trap: Bullies Win Every Argument about the Project and Product

In *Re-Evaluate the Value of a Bully* on page 161 we showed an example of how bullies can destroy a distributed team. Bullies might be well-meaning experts who want to do the "right" thing for the product and the team.

As a leader, consider how you recognize potential bullies:

- They drive everyone else's work.
- They become a bottleneck for everyone else's work.
- They—and only they—define the team's culture for starting and completing the work.
- They "correct" other people's work without notice or permission.

These people often mean well for the product. However, these people are incongruent. They think of themselves, and often the product. They might not think about the other person or the context. See *Assume Good Intention* on page 29 for more information.

How can you, as a distributed team leader—regardless of your title—catch bullies before their behavior destroys cross-team collaboration and respect? Consider these options:

- Encourage everyone to notice incongruent language, especially in themselves. Sometimes, it's easier to notice incongruence in others. In that case, consider offering feedback when you see incongruence. We also like to ask for feedback when other people notice our incongruence.
- Sometimes, people are unaware of the effects of their language. We've noticed this when people move from collocated to distributed teams. What might be acceptable in person—because of their tone of voice, facial expressions, and physical movements, not just the words they say—is unacceptable at distance. In this case, offer coaching or ask the person to find a coach to help with this change in their work style.
- Go meta and look for the root causes of why people bully others. Often, their behavior—especially if it's new behavior for that person—is a result of some sort of pressure. We all exist within an organizational system. Most often, our behavior reflects that organizational system with all its pressures. See an example of collocated bullies in *Retrain Your Code Czar.*[4]

---

[4] https://www.jrothman.com/articles/1999/01/retrain-your-code-czar/

- Depending on what else you notice, decide if it is time to bring the team together to build respect in person.

Whatever you do, don't ignore bullying behavior. Bullies destroy the safety, teamwork, and respect an agile team requires for collaboration.

8.12.5 Trap: Assume Vendors Are not Part of Your Project or Program

We often think that our projects or programs are self-contained inside our organization. However, if your project or program requires help or components from a vendor, it's worth your time to build a respectful working relationship with that vendor. Especially since—by definition—your vendor is not collocated with your team.

Consider these options:

- Visit with the vendor and/or ask the vendor to visit with you, to build understanding and respect for the requests each of you has of the other. We've seen agile teams ask vendors for biweekly "drops" of interim product, when the vendor had never used that kind of approach before.
- If you have a contract, discover the people behind the contract so you can build relationships with them.
- If the vendor is providing development support in the form of developers and testers, invite that vendor-developer to all relevant team and company meetings, including planning, retrospectives, hackathons, and face-to-face meetups.

8.13 Now Try This

1. Ask the team to discover each other's preferences using the *Compass Activity for Distributed Teams* on page 257 and their team values through Identify the *Team's Values* on page 154.
2. Develop working agreements with the team once they understand their preferences and values.

3. Visualize *when* everyone works or is available for collaboration so the team members can connect and build empathy and respect.

4. Assess how safe *you* feel with your team to discuss issues, encourage learning, admit when you don't know, acknowledge problems, and set boundaries for each of your decisions. Are you and your team safe enough? What can you do to increase safety for everyone? That might be something you consider discussing with the team. Start with one-on-one conversations and then open it up with "learning conversation" with the team.

Now that you understand how you amplify empathy to build working agreements for your distributed team, it's time to think about how they can adapt their agile practices.

CHAPTER 9

Adapt Practices for Distributed Agile Teams

When we work with geographically distributed agile teams, their first questions are often about tools and practices. That order may make sense, until you think about the problems of distributed agile teams.

Collocated agile tools and practices do not translate directly into the distributed environment.

9.1 Identify the Principles Behind Your Potential Agile Practices

The agile principles suggest agile practices for collocated teams based on their context. Use the agile principles to decide how to create practices for your team: the type of team you have, the team's culture, and the overall organization's culture.

You may find that some practices translate and some do not. Adopt an experimentation mindset: you may need to experiment with or develop new practices based on principles for successful distributed agile work.

 Understand your distributed team's system of work before tackling any tools or practices.

We don't recommend you take any existing collocated team agile practice and "transfer" it to your distributed agile team. We see too many practice failures.

Start with understanding the principles behind a given practice before adopting any practice. Is the principle about team collaboration, customer collaboration, or double-loop learning? Use that understanding to decide what to keep, remove, or change for your distributed environment.

The retrospective is the very first practice—a "tool"—any agile teams needs. And, not just one retrospective, but a cadence of retrospectives. The retrospectives, with their output of experiments, builds the team's habit of regularly improving how they work together and deliver value. The retrospective provides teams a chance to examine their assumptions, as well as their work—double-loop learning.

9.2 Reflect Often as a Team

The retrospective is the one necessary meeting for any agile team. It does not matter what kind of a team you have, or the team locations or timezones. Agile teams look for ways to inspect and adapt. Stop doing that, and your team will no longer improve how they deliver value.

Assuming your team has sufficient hours of overlap, consider a synchronous and collaborative retrospective. If your team does not have sufficient hours of overlap, consider a retrospective where a portion is asynchronous (e.g., data gathering) and another portion is synchronous (discovery and decision).

Every team member, including the facilitator, will need to adapt how they reflect. The team will adapt to the communication and collaboration tools available. The facilitator also adapts their facilitation skills and how they use—and invite the team to use—the communication and collaboration tools.

9.2.1 *Consider Your Retrospective Approach*

We happen to like the retrospective framework from *Agile Retrospectives: Making Good Teams Great*, DER06. Consider which parts need to be synchronous:

1. **Set the stage**: Can initially be done in email or a text backchannel. Once the team actually meets synchronously, consider

how you can bring "all the voices into the room," as in *Agile Retrospectives*, DER06. Watch for safety based on who participates in any synchronous meeting. For example, too many distributed teams do *not* see the Product Owner as a part of the team.

2. **Gather data**: Can start solo and asynchronously. Each team member gathers their data asynchronously in advance. At the very least, when the team meets, they can review the data together to gain agreement on the data.

3. **Generate insights**. Best done synchronously. This is tricky and often requires an experienced facilitator to help the team move through the "Groan Zone" (as in *A Facilitator's Guide to Participatory Decision-Making*, KAN14). That's the time where the team sees its challenges and has too many disparate opinions about what to do. Use video for this part of the meeting, to be able to see everyone's body language.

4. **Decide what to do**. Best done synchronously. Many teams have plenty of ideas for solving their issues. The problem is narrowing down those ideas to create small experiments. Many teams let these experiments run for months, so the facilitator may need to ask the team to finish the experiment and decide.

5. **Close the retrospective**. Best done synchronously. This is where the facilitator confirms that the retrospective helped the team discover what they should amplify (e.g., doing well) and what they might change. If the participants didn't feel the retrospective was effective enough, gather suggestions for improvement synchronously—if time permits—or asynchronously.

9.2.2 *Use Video for Your Retrospectives*

Even if your team does not use video for other meetings, insist on video for the distributed teams' retrospectives.

When we see each other, we learn together. We learn other people's reactions and how they clarify issues. When team members can't see each other, they miss the facial expressions and body language. Team members discover their strongly-held beliefs when they see each other. Without video, people often misinterpret other people's statements.

Sometimes, video helps people create psychological safety because video offers rich and natural communication. Other times, people feel exposed when they use video. When any team member does not feel safe, the value of the retrospective to the entire team diminishes. Then, team members will raise only minor problems, if any.

Some people say they feel safer with video off. In our experience, that's a false sense of safety. When people don't use video, they often feel "anonymous" and can lob verbal grenades into the meeting. How can you, as a leader, help everyone build safety so they can discuss anything, without needing the false safety of no video?

See if you can discover a person's discomfort in advance of the retrospective. If not, do attempt to discover the discomfort afterwards. Often, the facilitator will follow up asynchronously, after the retrospective. We will not address the fine points of facilitating a distributed retrospective. That's a different book.

9.2.3 *Use a Temperature Reading When Emotions Are High*

Sometimes, a standard retrospective isn't exactly the right tool to understand how people feel about an issue. When teams struggle with the work, the format of "what works, what doesn't" may not get to the heart of the problem. In that case, consider a temperature reading. This approach is modeled after the work of Virginia Satir.

In this format, the team examines the recent work through the following questions in this order:

1. Appreciations. An appreciation is a personalized thank you with the reason why. It takes the form of "(*First Name*), I appreciate you for (*something*) because (*for some reason*)."
2. New Information. What new information does anyone have that others may need?

3. Puzzles. What puzzles you?
4. Requests for Change. What changes would be helpful? Sometimes, people know they want *something* to change, and don't know what that something is. In that case, they can ask the team for help to articulate the request for change.
5. Hopes and Wishes. What hopes and wishes do you have?

We have found that the Temperature Reading helps people discover and manage their emotions in a constructive way. Many of us feel strongly about our work. We want to feel proud of what we do and how we do it. Or we may be angry about being blocked in our work or saddened that the team was unsuccessful in delivering the planned value. And, we don't always agree about how to proceed.

Use the Temperature Reading to surface and navigate these scenarios with maximum respect for the people.

9.2.4 *Verify Your Team's Focus: Urgent or Important?*

If your team performs retrospectives on a regular basis, you might discover the people and the team are stuck in urgent thinking over important work.

Stephen Covey in *The 7 Habits of Highly Effective People*, (COV89), discussed the difference between urgent and important work. The idea is we know we have important work, such as building respect and safety in the team, as well as periodic reflection. The problem is that other work is so important—so urgent—that we think we can take shortcuts and "just get to the work." Covey referred to this as "put first things first."

The problem is this: We are human. When we take a little time to build safety, respect, and transparency (see *Create Transparency at All Levels* on page 25), we create a team that works well together. The team can discover ways to address difficult-to-discuss issues. The team doesn't get stuck on the work or how to improve the work. They also learn to respond quickly to surprises in the work because they already established safety.

Urgency poses a unique threat to psychological safety in distributed teams. When teams have an impending deadline, they feel an even stronger sense of urgency. That urgency translates into focusing on delivering value over any other team activity, including retrospectives.

Collocated teams may also feel that urgency, and everyone can *see* them working. The deadline feels more urgent for distributed teams because they are literally invisible. Does each member of the team recognize everyone else's contribution?

Even without an impending milestone, team members might feel that the retrospective isn't as important as monitoring email or watching server logs, or squeezing in one more online code review. When a distributed team misses a retrospective, that could be the start of a downward spiral.

Even if the distributed team spends the time in a retrospective, they may not feel the necessary focus to inspect and adapt their work.

When teams feel this urgency, they need the support of a servant leader, the more senior the better. The leader explains that the retrospective is vital to the work of the team. By taking the time to pause and reflect on the goals and work of the team, they can deliver even more value than merely turning their regular work crank.

9.2.5 Consider Kaizen

Some teams don't perform retrospectives as a separate event. Instead, a kaizen event becomes part of their weekly cadence to provide a more frequent inspect-and-adapt point for the team. Some non-collocated teams may add this to the end of a daily stand-up or wall walk.

Shorter Kaizens for Knowledge Work

Although they are both called "kaizen" events, a kaizen event for a team of knowledge workers is not the same as a kaizen event for a manufacturing team. Knowledge workers can use an hour-long

> reflection every week or two. Manufacturing teams might need a
> week-long event every few months.

The team gathers issues to discuss on an ongoing basis and retains them using some type of tool, such as a chat channel, a wiki page, or a column on their board. They keep the issues visible as they proceed throughout the week.They might have a way of ranking these issues that might be as simple as "How many people feel strongly about this issue?"

The team uses a 30- to 60-minute timebox to discuss the issues, select alternatives for experiments, and create action plans.

When teams use kaizen on a more continuous basis, they don't need the ceremony of the retrospective as often. They tend to catch issues when they are small, rather than large. These teams may have a retrospective on a specific theme or focus to examine and set an experiment around a larger issue.

Once you've helped the team understand how to reflect and use those observations for improvement, think about the kind of rhythm your team needs.

9.3 Create the Team's Rhythm

In *Create a Project Rhythm* on page 32, we recommended your team have a rhythm, a cadence to its work. You can use timeboxes as a cadence. You can use flow and still have a cadence for the team's planning, story creation, delivery, and retrospectives.

When teams use Scrum or other iteration-based agile approaches, the timebox creates the cadence. The timebox defines when the team plans—at the start of the timebox. The timebox defines when the team demonstrates—at the end of the timebox. Many teams workshop stories in the middle of the timebox, especially if the team uses a two-week timebox. While teams can deliver all the way through a timebox, the timebox creates an expectation that the team will deliver *at least* at the end of the timebox.

Your team can also use flow, with a cadence of delivery, planning, and reflection. Your team might decide to demonstrate and plan once every two weeks. The team might decide to plan whenever they complete two or three stories, not once in a timebox.

Teams that use a flow-based agile approach will decide which project events they need at what kind of a cadence. These events might be meetings to walk the board, workshop stories, or other meetings the team needs.

When a team—especially a distributed team—finds their flow, they might be loathe to interrupt their flow. Even if it's the right cadence to have a project event, they are in flow and want to stay that way.

Build and Maintain Momentum

One iteration-based agile team had discovered their flow. They had six to seven hours of overlap, depending on the day. They managed the WIP by collaborating on stories as pairs or triads. They used the timeboxes as a cadence for their planning and reflection.

One Tuesday morning, they were supposed to workshop stories with the Product Owner. However, the team members felt that both open stories were very close to being done. They interrupted themselves to workshop stories, and decided to ask the question in their retrospective: if they were "very close" to being done on their stories, should they wait to workshop more stories?

They decided to run an experiment the next iteration. They deliberately made their stories smaller, so they had a better chance of finishing them and not being in the middle of a story when it was time for more planning. However, one of the stories proved to be much larger than they expected.

On that next Tuesday morning, they not only postponed their workshopping, but asked everyone on the team to collaborate to complete this large story. They finished that story two days later.

> The PO worked with the team to refine the "story" as they proceeded. The PO thought they needed it "all" to release. The story turned into a feature set, instead of one story. The PO sat with the team, refining the paths, which became new stories, and the acceptance criteria for each path.
>
> They used that experience to inform the next batch of story writing. What had made this story so large? How had they missed that?
>
> They discovered many areas of complexity in the code and lack of tests. When they decided to maintain momentum, they were able to avoid repeating the story vs. feature set confusion.

Break your team's rhythm with caution.

Maintaining the same cadence or timeboxes indefinitely might not work for your team. The team needs to pay attention to the kind of work it has *now*, not the kind of work it had *then*. See *Select Iterations or Flow* on page 111.

Use your experimentation mindset to decide how to create your team's rhythm, and when it's okay to break that rhythm.

9.4 Consider Which Meetings You Need and When

In addition to retrospectives, agile teams often need to meet to perform this work:

- Plan enough for the next chunk of time or work.
- Review the progress of that work (not the status).
- Understand how the team can collaborate to move that work to done.
- Workshop stories/understand work so the team can plan for it in the near future.
- Gather product-based feedback.
- Adapt your product planning to accommodate your distribution.

Your team might need other meetings that we haven't discussed. For example, people might work together as subsets of the team when they spike or perform some preparatory research. We prefer an all-team approach to a spike, and we realize that some distributed teams can't work together because of their hours of overlap.

Depending on the hours of overlap, teams will need to be judicious about which collaboration events they need and how the the team organizes those events.

Think about events in these ways: keep the team moving forward and setting new directions:

- Chartering a new project
- Recreating a team
- Team-based planning for longer than two or three months
- [a] Strategic planning for a program or product. We won't discuss that in this book.

Sometimes, a technical issue can instigate a need for a meeting.

Break the Email Chain

One team liked to debate issues in email. They would continue in long email chains with replies going to everyone on the team and, as they continued, included broader email distribution lists.

As the company grew, many technical leads and managers realized the emails wasted everyone's time. One of the senior managers recommended breaking the email chain if the issue had more than two long replies.

This team had sufficient hours of overlap that they could schedule a meeting with enough of the interested parties. They timeboxed the agenda and resolved the issue in the meeting. They made reasonable decisions and sent out the summary of the rationale and the decisions.

Your team might not have sufficient hours of overlap to break the email chain with a synchronous meeting. In that case, see if the team members will accept a recommendation from a trusted third party. Or, ask the team members to create experiments and try the two or three competing solutions.

Whatever you decide, realize that long email chains indicate a need for rich and natural communication between the team members.

The meetings in this chapter help the team continue to move forward. Let's start with planning, so the team knows what they will consider for the next short time period.

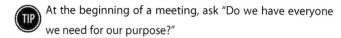 At the beginning of a meeting, ask "Do we have everyone we need for our purpose?"

Once you know which meetings you need, make sure everyone can participate.

9.4.1 *Plan as a Team*

Agile teams have two parts to the planning: what they will do, and when they expect to do it. When teams workshop stories, they define the "what" they will do. When teams plan for an iteration, that's the "when."

Back in *Select Iterations or Flow* on page 111, we recommended your team select an agile approach that fit their hours of overlap: iterations usually work well when a team has more hours of overlap, while a flow-based approach might be better for teams with low hours of overlap.

If the team has sufficient hours of overlap to plan as a team, and you are not sure about agile approaches to planning, consider these resources:

- *Agile Estimating and Planning*, COH05
- *Create Your Successful Agile Project*, ROT17
- *Collaboration Explained*, TAB06

Some teams have no hours of overlap. In that case, it's very difficult to plan as a team. Consider these options:

- Use a copilot *Consider a Copilot or Proxy Facilitator* on page 140 to help facilitate the planning. This option asks much of the team to time-shift for the synchronous meeting.
- Some team members might voluntarily time-shift to plan a little until the team members can all meet in one location to replan together.
- Some of the team could plan *for* the entire team. That's not an agile approach. It breaks the agile principle of collaboration, and our principle of *Default to Collaborative Work* on page 36.

Planning is not just about deciding which work to do when. Planning is also a learning activity. When some people aren't there or aren't awake, the team members don't learn the same thing.

You've noticed we insist on the team planning together. We insist, not for collaboration sake. It's because *without* the collaboration, your team can't be an agile team.

The whole point of an agile team is to collaborate within the team to learn and to deliver. The agile team collaborates with its customer or via the Product Owner to learn about what the customer wants, why they need it, and how to deliver what the customer needs.

If your team can't collaborate in this essential activity, can you really be an agile team?

If you think your team can't really live the agile values, consider the options in *When Agile Approaches Are Not Right For You* on page 17.

9.4.2 Review the Work in Progress

Many collocated teams use standups to see the state of the work, the impediments to continuing the work, and to generate team-based measures. Flow-based teams also see where the work is stuck.

Flow-based teams often call this "walking the board" or "wall walks," rather than a standup.

Regardless of what you call this meeting, distributed teams need to see their work and the state of all the work.

We like these questions to review the work state:

- What do we, as a team, need to do to move this work to done?
- Do we have any hidden work, work that's not on the board?
- Do we have work that's waiting for something or someone?

Do not conduct standups if your team has insufficient hours of overlap. Those kinds of meetings are not worth the aggravation. The team can't work as a team if they can't use a standup to micro-commit to each other.

You might need to consider the options in *When Agile Approaches Are Not Right For You* on page 17.

Beware of walking the board asking about individual work. Distributed agile teams collaborate on the work.

9.4.3 *Collaborate on the Work to Move it to Done*

Too many people assume that distributed teams can't swarm, pair, or mob on the work. Not true.

All of the collaboration practices depend on sufficient hours of overlap. It's not the distribution that is a problem—it's the availability of other people to work together.

Swarming is when the entire team works together on one item until that item is complete. The team decides how to approach the problem. Each person works according to his or her primary technical skills. The team agrees when to check in with each other, as a team. We recommend about once an hour.

When one person finishes the work they can do, they offer to help someone else who is still working.

Pairing is when two people work on one item, with one virtual keyboard. The two of us pair-wrote this book. Distributed pairing can work if the tools allow it. The people in the pair need to be able to pass the virtual keyboard back and forth to each other often, at least every ten minutes. If one person has to only look at the other person's screen, pairing feels uncomfortable and uninviting to the person not typing.

Mobbing is when the entire team works on one item with one virtual keyboard. When mobbing, people pass the keyboard even more often, possibly every five minutes. If you wonder about how it's possible to mob as a distributed or dispersed team, read an agile team's mob experience report.[1]

One common case is to see where other team members may need help with testing or peer reviews. Some team members may prefer to do this at the end of their day. Others may wish to take more regular breaks and be willing to spend an extra thirty to sixty minutes helping team members with tests and reviews. Regardless, the team should have the discussion about these preferences and record their working agreement.

What Kinds of Reviews Does Your Team Need?

Some agile teams use code reviews instead of pairing or mobbing, regardless of their hours of overlap. Some agile teams need design or architecture reviews because of their product.

Teams review work products for at least these reasons: to help other people on the team learn about the work, and to create a product with integrity.

We are fond of pairing or mobbing to build in the review as the team proceeds. If your team has too few hours of overlap to pair or mob, your team will need to have asynchronous reviews. With asynchronous learning, it's even more important that your team decide, possibly as part of their working agreements, about the kind of reviews they need to move the work to done.

---

[1] https://cucumber.io/blog/2018/06/20/inclusive-benefits-of-mob-programming

In collocated teams, you can walk by someone else's office and see if they are available to chat. In distributed teams, you can't do that.

Instead, as part of the working agreements, decide how to use overlap time. Are there times during the hours of overlap that team members will:

- Ask for specific help
- Ask to pair or mob
- Ask for reviews (code or test)
- Ask for help testing (even if you have automated testing, humans are still better for many edge case tests)

When team members know it's okay to check in with each other, they can collaborate to complete work.

9.4.4 *Workshop Stories as a Team*

Distributed teams also need to workshop their stories together.

 The team loses the possibility of learning and clarification if someone else workshops stories and hands the stories to the team.

Distributed teams have many choices for how to workshop their stories:

- Have a synchronous meeting to workshop stories together.
- Use a combination of synchronous and asynchronous work. Clarify the intent of the story (or stories) in a synchronous meeting. Then, everyone writes "their" stories, possibly with one or two other people. Gather and review the stories in another synchronous meeting. This works when the team has fewer hours of overlap.
- Consider who the product representative is and how that person can work with the team. We have seen business analysts work with Product Owners—especially overworked

Product Owners—who can then represent the Product Owner perspective to the team.

An agile team needs to understand the stories they have in the near future. When a team workshops the stories together, they develop a shared understanding of the work. They have a pretty good idea if this is a feature set or one story. They might even realize that the work is right-sized for the team or too large.

9.4.5 Gather Product-Based Feedback

Teams need to show themselves and their stakeholders their product progress on a regular cadence, via a demo. Where are your stakeholders located?

Many times, the teams will be asked to time-shift to demonstrate their work at a time convenient to the stakeholders. While this may work if stakeholders are all in one time zone, it can be extremely challenging for the team if stakeholders are spread across multiple time zones.

When stakeholders are spread across time zones, the team has some options. The team may choose to do a limited number of short review sessions to demonstrate the work and gather feedback.

Another alternative is for team members to record videos of their demonstrations and distribute them two to three days in advance of a review meeting where the team can answer questions in real time. Ask stakeholders to provide feedback via email, chat, or a dedicated feedback page where all stakeholders can see each other's comments.

When all the stakeholders can see each others' comments, they can clarify their comments and play the perfection game—what would make this feature even more perfect?

Product feedback is different from project-based and product-based measures. Distributed teams have the same needs for quantitative and qualitative measures as collocated teams do. See *Create Your Successful Agile Project*, ROT17 for details.

9.4.6 Adapt Your Product Planning for Your Distribution

The more distributed your organization is, the more you might need a whole-organization product value team.[2] That whole-organization team creates a larger product perspective

In collocated teams it might be possible for your product champion or product manager to meet with other relevant Product Owners on a regular basis. Those people might meet with the salespeople on a regular basis. The product people can plan the roadmap and organize the near and far backlogs for the teams.

Sometimes, the salespeople need to talk directly with the technical teams to explain the customers' problems. Sometimes, the support people need to talk with the salespeople and the technical teams to explain what they see.

We are not suggesting a free-for-all for product planning. We are suggesting that a tight hold on the product planning doesn't fit for a highly dispersed organization. We recommend you have a regular cadence of planning, gathering information as you proceed from all the people with customer or problem information.

Consider experimenting with your product planning to take advantage of all the people with customer and problem information.

When people work with others they understand and trust, they are more likely to be successful as a team, regardless of location.

9.5 See Your Agile Practice Traps

Distributed agile teams encounter plenty of agile practice traps. We've outlined some ways teams can adapt the collocated practices to work with a distributed agile team. Always ask yourself if you need to adapt a specific practice to make that practice work for your team.

---

[2] https://www.jrothman.com/mpd/2017/09/alternatives-for-agile-and-lean-roadmapping-part-5-the-product-value-team/

In addition, we see too many of these agile practice traps:

- The Product Owner has no hours of overlap.
- The team doesn't own their agile approach.
- Distributed retrospectives run the same way as collocated.
- The team focuses on the urgent over the important.
- People work as individuals.
- Teams meet with insufficient hours of overlap.

Note that the agile practice traps stem from a common cause: insufficient hours of overlap.

9.5.1 *Trap: Product Owner Has No Hours of Overlap*

The Product Owner is an integral part of an agile team. The Product Owner provides the understanding of why each feature brings value to the customers and the business. The Product Owner therefore needs to set priorities for the team to deliver the most business value as early and often as possible.

We see too many teams whose members have insufficient hours of overlap with the Product Owner. That means the team members—possibly the entire team—cannot plan together, at all. The Product Owner isn't available to answer questions, clarify the acceptance criteria, or accept a story. Never mind see or run a demo.

It's a disaster.

We don't recommend you try an agile approach if your Product Owner has no hours of overlap with the majority of the team.

Consider ideas in *Follow the Sun* on page 44 to see what you can do. Map the value stream as in *Map the Value Stream to Visualize Cycle Time* on page 46 to see who waits for which kind of work. Also see *When Agile Approaches Are Not Right For You* on page 17.

9.5.2 *Trap: The Team Doesn't Own Their Agile Approach*

Too many distributed teams don't own their agile approach. That means they don't:

- Make changes based on a retrospective or purposeful experiments.
- Set a cadence to review progress with stakeholders.
- Overlap with the PO in working hours.

Distributed agile teams may start from a "standard" agile approach (a well-known framework) to help them learn how to adapt to an agile way of working. Sometimes, even with frequent retrospectives, the team's progress seems slow.

Managers may be tempted to change the team's process to help the team keep moving. However, without a retrospective to gather everyone's perspective and generate insights about the actual challenges of the work, the manager might instigate a wrong change.

When the team decides on a change, help the team create an experiment with a defined goal and duration to check in on the change. If it's successful, then the change can become part of the team's process. If not, it's time to create a new experiment.

It's the same problem with setting a cadence to review progress, as a team and with stakeholders. If someone outside the team sets the review cadence, the team members feel as if "agile" is telling them how to work. That's not the point of an agile approach.

Agile approaches provide frequent feedback to the team and stakeholders because the team deliver small iterations of value. If the team can't deliver to a frequent-enough cadence, the team needs to understand and solve that problem.

Once the team can deliver, the stakeholders need to be available to see what the team completed. That means the stakeholders have to have sufficient hours of overlap with whoever provides the demonstrations.

If the team doesn't own their process—if someone else owns their agile process—they will never acknowledge or solve their big problems.

9.5.3 Trap: Distributed Retrospectives Run the Same Way

Often, we discover distributed agile teams that skip retrospectives or hold short ineffective retrospectives. Their reason? Being distributed makes it difficult to have a good retrospective. We don't buy this.

Distributed retrospectives require a facilitation mindshift. The intent behind the facilitation is the same—to create a safe environment for the participants to inspect and adapt. How the facilitator and participants manage themselves change.

Collocated retrospectives use tactile cues and various in-person collaboration tools and techniques to create safety. For instance, if someone is dominating a conversation in room, the facilitator may quietly walk behind them and tap them on the shoulder.

While you can't literally tap someone on the shoulder, you might use a private message or a team agreement about how to manage a dominating person. Distributed retrospectives use agreements to create safety.

We've suggested several options in this chapter. While you might still need to adapt our suggestions, consider how much each person can equally contribute asynchronously and how much synchronous time the team needs.

Consider these questions:

- How much data gathering does the team need? How much of it is individual and could be done asynchronously? How much problem discovery time does the team need? How much of that is asynchronous or preparatory, and how much is synchronous?
- How much time does the team need to select the one thing they might experiment with next?
- How much time does the team need to create an action plan for their experiment?

These questions challenge many distributed agile teams' retrospective practices. That's why the retrospectives are so difficult. If your team has trouble reflecting at a distance, consider the ideas in *Non-Collocated Teams Deserve Face-to-Face Time* on page 12.

Agile teams who *use* their retrospectives to strengthen their process discover several benefits.

- When teams discuss what works and what doesn't work, they can create experiments and refine their approach, as a team.
- They also establish psychological safety for team members to speak freely about problems and look critically at their work to engage in double-loop learning.
- The retrospective helps the team focus on their team goals to discover and decrease the causes of misunderstandings.

The team members feel safe to question anything to improve the work of the team.

If your team's retrospectives aren't working—if the team isn't discovering and fixing problems—set aside team time to discuss that issue.

9.5.4 Trap: The Team Focuses on the Urgent Over the Important

Back in *Verify Your Team's Focus: Urgent or Important?* on page 177, we discussed possible issues of urgent over important.

Urgency increases when we don't have enough hours of overlap.

Collocated teams are each other's work "family." They are literally in each other's physical space for eight (or so) hours a day. They share a system of work and understand each other's work.

Distributed teams—even the geo-fence team members—are not in each other's physical space. Distributed team members have to create a shared system of work and learn about each other's work. When their hours do overlap, they may feel urgency in accomplishing tasks, but not necessarily for improving their process or strengthening the team.

If your team is still feeling pressured by the urgent, instead of being proactive with the important, consider these options:

- Conduct one-on-ones with each person to understand their perspective. You might have to time-shift to do so.
- If you have not used the *Compass Activity for Distributed Teams* on page 257, consider taking the time to do so now.

- If the team has not yet taken the time to *Create Working Agreements* on page 156, do that now.
- Learn more about the impending deadlines, and see if you can ease them.

Help your team escape the tyranny of the urgent, so they can focus on the important.

9.5.5 Trap: People Work as Individuals

Especially with insufficient hours of overlap, teams (and their managers) fall into resource efficiency thinking. Instead, *Think in Flow Efficiency* on page 10.

This chapter is about how to adapt the common collocated collaborative practices to your distributed team. Do not think that a "team" composed of individuals can be an agile team. See *Default to Collaborative Work* on page 36 for why your team needs to collaborate.

If your team members cannot collaborate, consider using another approach for your project. See *When Agile Approaches Are Not Right For You* on page 17.

If your team has insufficient hours of overlap, review the chapter *Avoid Chaos with Insufficient Hours of Overlap* on page 41. Your team might not be able to collaborate and work as a team.

9.5.6 Trap: Teams Meet with Insufficient Hours of Overlap

The fewer hours of overlap your team members have, the more difficult it is to manage meetings.

Consider how you can have meetings without sufficient hours of overlap. Can you:

- Alternate meeting times so everyone shares the time-shifting meeting pain?
- Set meeting times so all team members are able to participate?

- Share information about the meeting content and decisions asynchronously if people can't participate?

We do recommend creating cross-functional teams at each location if you have too few hours of overlap. See *Trap: We Can Hire Experts Anywhere* on page 221.

Do not expect people to time-shift on a regular basis. That's not reasonable and will create problems in the work, not just in the meetings. When people are tired, they make mistakes. Planning mistakes (backlog creation and refinement) multiply into mistakes in coding and testing. Mistakes in micro-status (standup or handoff) create mistakes in project status.

Consider how you can help the team create a culture of asynchronous information-sharing to supplement the meetings:

- Does the team use a backchannel regularly to coordinate and to share tacit information?
- Does the team capture decisions and working agreements about how they work in some team workspace (e.g., a wiki, for instance, that anyone on the team can update)?
- Does the team document architectural and design decisions for all parts of the product in (UI, code, tests, database) in the team workspace?

The more information the team shares with each other on a regular basis, the more likely the team will work together as a team.

9.6 Now Try This

As you've seen, agile practices for distributed teams mirror those for collocated teams. However, agile practices differ for distributed teams.

1. If your team has insufficient hours of overlap, how will you manage your team's meetings? What choices can you create?
2. Consider how your team will reflect, to be able to inspect and adapt to experiment and improve their agile approach.

3. What kind of a rhythm does your team need to plan, deliver, demonstrate, and reflect?

4. How can you make the work and data visible to others outside the team without requiring a synchronous status meeting?

Now that you've seen how to adapt agile practices for your distributed team, it's time to see how to add or change people on your distributed agile team.

CHAPTER 10

Integrate New People Into Your Distributed Agile Team

Distributed hiring activities challenge even the most experienced human resource staff and agile teams. Too many candidates talk about agile approaches but do not practice the agile mindset or have successful agile experience. You think you've got someone good and then you interview them. You realize they aren't going to work because they don't fit your agile culture or are not familiar with how to work well remotely.

In this chapter, we will only discuss the specific hiring and integration issues for *distributed* agile teams. You still have the challenges of ensuring you look for diversity (such as race, gender, work preferences, and various experience), hiring for cultural fit, maintaining a hiring pipeline, and preparing properly for interviewing and integrating new team members.

Collocated teams have an advantage over distributed teams because everyone can see each other. The candidate can see the team and the work environment when the candidate has an in-person interview. The team can see the candidate and the candidate's reactions.

Distributed agile teams also encounter the "How agile is this candidate?" question. In addition, because the candidate doesn't meet the rest of the people in person, the hiring team might have trouble assessing how well the candidate will fit the team's culture, and whether the candidate will fit the organizational context.

We'll address how to manage those challenges in this chapter.

To manage the fit and context questions, distributed agile teams require people who are:

- Able to take more initiative in the work
- Able to create good working relationships in the team
- Able to effectively work in a distributed environment
- Passionate about the work
- Willing to continually inspect and adapt their approaches to the work

When people are passionate about the work, they continually adapt to the work environment (both the agile part and the distributed part). This adaptation requires extra initiative by team members and the team overall. They need to "own" their environment—the work and the working relationships. These qualities allow them to be successful in a distributed environment.

While technical skills do matter if people are distributed, interpersonal skills make or break the team.

10.1 Focus on Interpersonal Skills

Interpersonal skills are the qualities, preferences, and non-technical skills that define us. Distributed teams amplify each person's presence or absence of interpersonal skills. Without constant attention to interpersonal skills, we tend to "focus on the work" over focusing on collaboration. What we do not see becomes easy to ignore and if we do not make an effort to see and hear our team members, our interpersonal skills degrade quickly online.

This is why *Non-Collocated Teams Deserve Face-to-Face Time* on page 12. They need this connection on a regular basis. Not only for setting large goals, but for making stronger connections on the team.

One of the most important connections to make starts with the initial integration of a new team member on the team. If they experience a team that mostly works in personal silos, they will follow that example. Contrast that experience with teams who regularly pair or mob or collaborate across distance and time. People follow the example the rest of the team provides.

Consider this approach to mapping the principles to interpersonal skills for a distributed agile team:

Principle	Interpersonal skills that reflect this principle
Acceptable hours of overlap	Empathy for other people's time and availability
Transparency at all levels	Self-regulation and transparency of own work
Culture of continuous improvement	Experimentation mindset, self-directed, high initiative
Pervasive communication	Able to explain state of their work and ask for help. Available for coaching
Assume good intention	Empathy for other people's competencies, work preferences, and normal and abnormal communication patterns
Create a project rhythm	Providing reliable results and feedback on a regular basis
Culture of resilience	Adaptability for own work, ready to adapt to fit the needs of the team
Default to collaborative work	Work for the betterment of the team, not individual accomplishments

Figure 10.1: Interpersonal Skills That Reflect Principles

Based on that, consider this general list of interpersonal skills:

- Empathy
- Able to manage own work and work environment
- Able to collaborate on the work, the process, and the team's learning
- Able to offer and receive feedback and generate results on a regular basis
- Adaptability
- Collaboration

Not everyone has or needs all of these skills to the same extent. Perform a job analysis[1] to define the essential and desirable skills for this role. That way, you can decide which skills are essential and which are desirable for your team.

Consider ways you can filter candidates initially on both a minimum technical skill set and their interpersonal skills. Technical people can learn new technologies relatively fast. If your team collaborates via pairing or mobbing, people can learn the technologies even faster. Then look for people who are capable of learning new technical skills fast, rather than looking for specific skills.

 Highly technical people who work as individuals can kill a distributed agile team.

Collocated agile teams can survive people who work as individuals. They have options for managing the team's interactions with that solo worker.

Distributed agile teams—if they want to use agile approaches—cannot survive a person who insists on working solo. The team will not succeed if all the team members are not willing to collaboratively experiment and solve problems together. The team loses its affiliation (container) and its shared knowledge, the differences become magnified, and the team members stop exchanging information. The "team" falls apart.

You might change how you recruit people if you want to find people with more interpersonal skills who can learn technical skills quickly.

10.2 Recruiting People for Your Distributed Agile Team

Many sourcing approaches remain the same for a distributed team as a collocated team.

---

[1] https://www.jrothman.com/htp/job-analysis/2013/08/are-you-overspecifying-your-open-jobs

We recommend the sourcing strategies in *Hiring Geeks That Fit*, ROT12. Ask your customers and current employees for referrals. In addition, ask your loose connections. We have found that loose connections (friends of friends/colleagues of colleagues) are more likely to introduce you to potential candidates. See The Power of a Loose Connection.[2]

There are three notable sourcing changes for a distributed team:

- How to express your job analysis without shorthand for the job description.
- When to hire from a customer.
- When to specifically look for people in various locations to take advantage of their local area.

10.2.1 Rethink Your Job Description

We often use shorthand or surrogate descriptions to discuss how to recruit people. For distributed teams, one of those surrogate descriptions is to say, "We need people who understand open-source development."

Instead, be specific in your job descriptions. Answer the question, "Why did we ask for this?" and use those answers in your job description. For example, instead of saying you want an open source developer, explain that you want someone who can follow and contribute to a team's process, who wants review of their work, and is open to collective code or test ownership. Use those words in your job descriptions.

While many open source people understand the value of transparency and collaboration at distance, not all of them do. We have seen some open source developers be unable to work in a collaborative fashion in real-time. Those people are not a good fit for a distributed agile team.

---

[2] https://www.jrothman.com/htp/agile-job-search/2012/04/the-power-of-a-loose-connection/

10.2.2 **When to Hire from a Customer**

If you sell to people all over the world, you might decide to hire people near various customer sites. As your business succeeds, some of your customers might ask you for a job. Or, you realize someone at a customer site is perfect for a specific open position.

 Read your vendor contract to understand when you can and cannot recruit from customer staff.

Be aware that if you hire the internal champion, you might lose that particular customer. However, you might find that the new hire becomes a champion for many other buyers, as a location-specific expert.

10.2.3 Recruit in Specific Geographical Areas

You might want to specifically create feature teams local to your customers. You might not want to hire from a customer, but create a virtual office where your team members can interact with some customers on a regular basis. You can hire smart people from all over the world if you create feature teams with sufficient hours of overlap.

You might decide to source people closer to the current members of your team who do not want to relocate or commute to your buildings (e.g., in satellite or cluster teams). Expand the idea of "local" hiring, to take advantage of people who might be part of a team with at least four hours of overlap.

You can create a more resilient workforce when you expand your idea of what is "local." Consider cluster or nebula teams that can collaborate across many cities and time zones. If we rethink "local" as the same or adjacent time zone as our other staff, we could establish cluster or nebula teams. With sufficient infrastructure, these teams could successfully collaborate and also not be as impacted by weather or culture differences.

Your organization *must* provide the infrastructure and policies to support high bandwidth remote collaboration.

10.3 **Define Your Hiring Process**

Within your constraints (Human Resources, HR, or organizational), each team needs to define its own hiring process. Distributed teams adapt the hiring process that many collocated agile teams use.

- Establish collaboration for interviewing
- Screen candidates
- Develop useful interview questions
- Organize interviews with an interview matrix for coverage
- Decide on a candidate (similar to collocated teams, so we won't discuss that here)
- Reflect on the interview process

Before you start interviewing, part of your job as the hiring organization is to establish collaboration with the candidate throughout the interview process.

10.3.1 *Establish a Collaborative Environment for Interviewing*

The interview process balances two necessary elements: help the candidate have the best interview experience and see how the candidate reacts and works in a distributed environment. To that end, consider the tools your candidate will need to conduct phone screens (or preferably video interviews), technical screens, and manage the interview process from their end.

Consider asking the candidate to fully participate in the scheduling and organization of the interview. This helps everyone understand how the candidate understands and manages the hours of overlap and possible communications infrastructure issues from the start.

Some organizations don't recommend *how* the candidate organizes the interviews. You might learn from how the candidate organizes the interviews. Does the candidate attempt to schedule all the interviews as early as possible? Or, does the candidate complete one interview and then start to schedule more? Or, does the candidate schedule each interview sequentially, after they complete each interview?

When candidates schedule their own interviews, you might discover something about their interest in your position. You might discover how they manage their work. All of this data is fodder for questions in the interview process.

 Provide guidance, not direction, so the the candidate can be successful.

Let the candidate's progress through the process be part of the interview.

Be aware that some candidates might not have access to the necessary scheduling and audio/visual tools at the start of the interview process.

Offer Options to Candidates

One distributed organization provides candidates with interviewer names and emails. In addition, they recommend a series of free video tools, such as Google Hangout or Skype. They also recommend scheduling tools for the candidate.

The candidate then selects the video tool and schedules the interviews.

We do recommend a full 60-minute slot for phone screens and interviews. You might not need to use the full 60 minutes, but you might encounter technical difficulties. The 60 minutes provides everyone with enough buffer time to succeed.

Distributed agile team members are self-managing. Learn about how this candidate works with a self-managing team in the interview while helping the candidate succeed.

Make sure you create various screens to filter candidates in a way that makes sense for your team. (See *Hiring Geeks That Fit*, ROT12 for more details.)

10.3.2 *Develop Useful Interview Questions*

Good interview questions provide learning opportunities for both the candidate and the interviewer. The interviewer learns about the candidate's experience and perspectives from the answers. The candidate learns about the interviewer's culture from the questions.

The best interview questions provide both interviewer and candidate enough answers to know whether to proceed.

You might want to start with some closed questions to establish the candidate's experience, such as:

- How long have you worked in a distributed way?
- Do you have experience with our technology stack?
- What collaboration tools have you used?
- What is the longest you have worked non-collocated? (Consider whether your candidate needs extensive remote experience. People with one or two months of remote work might have different experiences than people with years of successful remote work.)
- How did the distribution work for you? Was it easier or more difficult for you to work distributed? (This is a start at more open-ended questions.)

As part of the interview process, you can check on the candidate's personal workspace. See *Virtual Work Requires a Reasonable Personal Workspace* on page 124.

You've both established that the candidate has useful distributed experience. Now, it's time to move to more in-depth behavior-description, open-ended questions, such as:

- Tell me about your team. (Was it a siloed team or cross-functional?)

- Tell me about how your team managed the distributed nature of work. (Look for team workspace issues as in *Communicate to Collaborate* on page 87 and *Create Your Collaborative Team Workspace* on page 111.)
- On your most recent project, what challenges did you face as part of a distributed team? (This question will let you discover whether the candidate has run into some of the challenges we have described in the book. They may not have the solutions, but if they are familiar with the issues and have some strategies for overcoming them, that can be positive for your team.)

People new to virtual work may not know how to deal with isolation and lack of collocated structure. You may need to explain why you ask these questions.

Explain the Why Behind These Questions

Both people can learn from the interview. The candidate may have a unique solution for a distributed collaboration challenge your team faces. Also, by telling the candidate why you have asked some of these questions, it helps them understand how you are looking out for your team's best interest.

One interviewer working with nebula teams says, "I'm asking these questions to understand if you know how to work without teammates sitting in the same physical space. That can be a big adjustment for some people and I want to make sure that you are comfortable with this way of working. I want to understand if you have considered whether our situation is sustainable for you."

You might want to say something more or less like this in an interview. It depends on the transparency you have in your team and organization.

As we wrote this section, we discussed how we might phrase these questions. We require video for natural and rich communication so we can see the candidate's reactions, especially if the candidate is new to remote work. Listen and look for surprise. If either the question or answer surprises either person, you may want to say more about *your* distributed work environment and how it is similar to or different from the candidate's distributed experience.

Now that you've established the basics of distributed team experience, it's time to dive deeper into the candidate's idea of an effective agile culture. You will start to see how this person may fit your distributed team's culture and a little about their interpersonal skills.

Back in *Cultivate Your Distributed Team's Agile Culture* on page 129, we suggested several areas that might create or reinforce a team's agile culture: sustainable pace, how the team manages its information, and servant leadership. You might use those areas to ask questions similar to these:

- Tell me about how you managed the pace of your work. Tell me about how your team managed the pace of its work. (See *Create Sustainable Pace* on page 133.)
- For the distributed teams you worked with, how did your team decide what information should be persistent and what might be just in the moment? (See *Manage Information Persistence as a Team* on page 135.)
- How did you and your teammates decide who would lead the development of specific features? Did you rotate this responsibility or was it delegated by a manager? (See *Watch for Unsustainable Help* on page 134.)
- How did your team discover and solve team-based problems?
- Have you coached, mentored, or taught others in your teams or organization? What was the need and how did you meet it? (See *Watch for Unsustainable Help* on page 134.)

You can't ask, "How empathetic are you?" Instead, ask about situations where you think you might see evidence of empathy, self-management, collaboration, etc. Here are some questions you might use to start:

- Have you ever been in a position where other people seemed to have trouble with the team dynamics? What was that? Did you act in some way? Describe your actions. (Pause and listen for the candidate's answer to each question. Note that the first three questions are closed questions, with specific answers. Only the "Describe" question is an open-ended behavior-description question.)
- Tell me about a time you had to create a new technical or virtual work environment so you could succeed with your team. What were the circumstances? What did you do?
- Tell me about a time you or your team appeared to have trouble collaborating on the work. What was the circumstance? What did you do? What did the team do?
- Tell me about a time you recently offered feedback as a remote person.
- When was the last time you received feedback in your distributed team? How was that feedback for you?
- How often do you personally deliver work? How often does your team deliver work? How did your distributed team measure the value of what they delivered? How did they monitor the work and the value delivered?
- Tell me about a time you had to change something about your work: the process, your specific deliverables, how you worked with the rest of the team, how you worked with a new stakeholder—something.
- What does collaboration look like for your team?

These questions are starting points. They allow you to set the context with the candidate and then ask for more details as the candidate explains his or her experience.

10.3.3 Organize Interviews with an Interview Matrix

Because your team isn't collocated, you won't be able to usher a candidate into the office, help the candidate move to various rooms, or bring the interviewers to that person. You'll need to change how to organize your team's interviewing.

Use an interview matrix to organize who will ask which questions when, using which kind of communication. Specify the communication method and acceptable hours of work time or time zone for a given interviewer. See *Hiring Geeks That Fit* ROT12, or the post, *Plan for an Interview with an Interview Matrix.*[3]

Also, as the candidate learns more about your organization, the candidate may change their answers slightly from interviewer to interviewer. A candidate who learns rapidly might be worth pursuing further.

Distributed interviews can challenge even the most experienced in-person interviewers. That's because of the need for rich and natural communication to build rapport. If you don't have both rich and natural communication, both the interviewer and the candidate might find it difficult to build rapport. If either person is unfamiliar with distributed work, they might not know how to create an environment where they have privacy and the ability to communicate.

People might be able to create a reasonable environment for a phone screen. They might be stuck for the interview.

In a collocated interview, candidates can literally see where people sit and how they might collaborate. Use the interview to describe how our distributed team works. When you paint the picture for a candidate, the candidate can visualize—or not— him or herself as part of the team.

---

[3] https://www.jrothman.com/htp/interview/2011/02/plan-for-an-interview-with-an-interview-matrix/

We've met a number of teams where they pair or mob with candidates to assess a candidate's skills. If you don't normally pair or mob, don't use the interview to pair or mob. You create the mistaken assumption that you do pair or mob on a regular basis.

In the same vein, if you normally work as a team, then a panel interview might work. However, we only recommend panel interviews if the candidate will work with several people at a time, such as a senior person or a coach. Even then, we recommend your first round of interviews be one-on-one, not a panel.

You might also need to think about who interviews which candidates. Select enough people from the team and include the manager. If you are interviewing a more senior candidate, or an agile "leader" of some sort (Product Owner, coach, Scrum Master, agile project manager), also include people from other teams. More senior candidates tend to influence and collaborate with people across the organization, not just those in one team.

10.3.4 Reflect Frequently on the Hiring Process

Distributed agile teams are even more self-managing than collocated agile teams because the managers cannot be a total part of the team. The team needs to make and live with their own decisions. That means the team needs to reflect on their hiring approach.

Does HR Facilitate or Infiltrate the Hiring Process?

Many organizations have an HR function that defines a team's hiring approach and constrains too many of the team decisions.

That doesn't work for agile teams and doesn't work at all for distributed agile teams.

HR exists to keep the company out of court. Do invite your HR people to comment on your job description, interview questions,

---

and your general approach. Ask if they can help you improve. Do use professional recruiters to help you find candidates who are diverse in experience, gender, nationality, and more. Diverse teams create great products.

Do *not* let your HR department insert themselves into your hiring approach in a way that sabotages your team's self-management. See *Hiring Geeks That Fit* ROT12 for more details.

Hiring is an iterative process. Make sure you assess and reflect on everything you do as you proceed, to continually improve it. You can make this reflection low-ceremony/easy engagement—a simple kaizen or a few questions are sufficient.

Here are some possibilities for your reflection:

- If candidates don't succeed in scheduling, assess your hours of overlap between the candidate and interview team members, and the tool choices.
- If candidates "fail" the technical screen, assess your filters.
- If candidates "fail" the interviews, assess your interview questions.
- If you are still unsure about a candidate, assess your interview matrix, questions, and interview team.
- If candidates reject your offer, assess your offer components or how the interviewers describe your company and the work.

When candidates accept your offer, you might want to reflect on your sourcing and how this candidate used your hiring process. Some distributed agile organizations conduct a hiring retrospective with new employees to discuss the process and how it worked for them. This gives them the candidate's perspective.

10.4 Use a Buddy System to Integrate New People

Consider asking the team to conduct one-on-ones with the new hire to create the human connection. That's helpful and not sufficient.

We also recommend a buddy system to help ease a new person's transition to the team, the product, the architecture, everything.

While a distributed buddy system is a bit different from the one in *Hiring Geeks That Fit*, ROT12, the intent is the same.

An integration buddy system uses one person as a full-time pair partner to integrate the person into "how we do things here."

When an existing team member dedicates a couple of weeks to collaborating with a new person, the new person collaborates and contributes much faster than if the new person works alone. They become integrated into the team.

We recommend each team member take an opportunity to be an integration buddy. Each person learns how their integration process might change, and each person gains a new perspective on how the product and the team works—and possible opportunities for improvement.

Your team might consider more traditional pairing, where people change pair-partners on a more regular basis, or even mobbing as a team. In our experience, the more the new hire works with other people on the team as a mob, the faster the team integrates the new hire.

As the new hire asks questions, the team learns what they should—or should not—change in their team process. Mobbing offers a fast feedback cycle for the team and the new hire.

Onboard or Integrate?

We use the word integrate purposefully.

Beware of the double meaning of onboard: too often, people think about onboarding a person into the company, not the team. People are not cargo to be loaded. This use of "onboard" hints at resource-efficiency thinking.

In contrast, when we integrate new people into a team, we think of them as resourceful, not resources. We think of them as having unique qualities, preferences, and non-technical skills that are essential for a successful team. Consider using the word "integrate" for your agile teams to build and maintain your agile culture.

Distributed people can't turn around or walk over to the person like they can in a collocated team. One option is to encourage the buddies to have a standing one-on-one for the first couple of weeks, to make sure the new hire is integrating into the team.

We prefer that instead of a one-on-one conversation, the buddy can request the new hire ask questions in the dedicated team backchannel, as in *Enhance Discussions with Dedicated Backchannels* on page 100. This becomes easier when the new hire meets one-on-one with each team member.

Asking questions of the team in a dedicated team backchannel helps the new team member and the team overall. Sometimes, a new team member's questions prompt the entire team's learning. We've seen this when the more senior team members realize they do not hold the same knowledge of the "best" answer to the question. Sometimes, this answer is based on different contexts.

The new hire's question helps recalibrate the team's knowledge. The new team member learns along with the team. Therefore, the buddy should tell the new team member how they help the team by asking their questions and remind the team to keep an eye out for these questions (see *Asking for Help Can Build Respect* on page 150.)

10.5 Plan Time to Integrate People

Every time you add or subtract a person on a team, the team changes. The team will have to rebuild trust and respect. If your team lost a person, that person might have facilitated certain team activities, provided unique skills, or provided informal support for part of the team's collaboration. Consider reworking the team's working agreements as in *Build Respect with Working Agreements* on page 147.

When teams integrate new members, the team and the new person experience a period of adjustment. Even people who move inside the company to different teams need integration time.

Teams Need Time to Adjust to New People

One existing team split into two teams, each with three veterans and two new-to-the-team people. The veterans said, "Everything is in the code. You just need to read the code."

The first month, the code reviews took longer than the team expected. The new people felt as if they weren't useful. The team hadn't used a buddy system, so the veterans didn't invest the time to help the new people understand the architecture and the code. Without investing this time in the team, the team took more time to complete the work. This team started to use a buddy system to help the new people learn faster.

Contrast that with a new team of just one veteran and mostly new-to-the-team people. The veteran coached and mentored new people on the team. Because there is only one veteran on the team, the team can't quite use a buddy system, but the veteran helps the new team members adjust to the code and the team. The veteran might have chosen to mob for more rapid learning, but in this case, the entire team didn't have sufficient hours of overlap.

Some veteran team members may feel conflicted by three goals: completing their work, guiding the work of the team, and helping new team members gain knowledge to contribute across the code base. In some cases, the "knowledge sharing" becomes a low priority. This approach does not support collaboration or a growing self-managed team. Instead, it reflects resource efficiency thinking (or pressure from others with this thinking). Consider asking the team to reflect on their three goals instead of seeing them as separate competitive activities.

Distributed teams might require more adjustment time than collocated teams due to the hours of overlap. Distributed teams might also need to be more deliberate in how they integrate new team members.

If there are fewer than four hours of overlap in a day, the new team member will only have brief opportunities for mentoring.

This makes it even more important to intentionally invest the time to build respect and understanding among all team members.

Distributed teams shouldn't "test" new people. However, because each person is unique, the team might need to revisit *Compass Activity for Distributed Teams* on page 257. The team will have to check that the new person can live with its working agreements—or the team will have to change its working agreements.

Teams don't need to change their project charter, and they *will* have to verify that the new person understands the project charter.

Sometimes the change to teams comes from growing the organization.

10.6 Scale Your Distributed Teams

The agile community uses the word "scale" in several ways. The first is to add new teams to the organization. The second is to add more structure so the teams can collaborate.

Let's first discuss how to add new teams.

10.6.1 Split to Scale Your Teams

Your teams might need to grow. As you add people to the teams, be aware that the team might become too large for effective communication. Too often, the natural and rich communication channels have trouble supporting more people on a call.

Larger teams often have a harder time with communication. That's because an agile collaborative team builds connections between each team member. (See *Consider Your Team Size* on page 65 for more details.) In that case, you might choose to split teams.

Nebula Teams Might Need Reconfiguration

Nebula team members have an extra challenge: how to encourage collaboration all the time. One large team had the problem of insufficient collaboration. People focused on the familiar, not becoming generalizing specialists—by choice—with the result that too few people understood enough of the code and test bases.

When the team realized this (via a retrospective), they decided to break into two smaller teams, one of four people and one of five people. They also decided to default to pairing or triads to work on their stories.

We also recommend you split a current team along tenure or experience lines, creating teams with a mix of experience to maintain and enhance your agile culture. The existing employees help instill the new employees with the desired culture. This provides cross-training for new team members, both for the product and the team's agile culture.

10.6.2 Scale Collaboration Across Teams

If you don't need to hire more people or create more teams, but need more teams to collaborate together, consider these scenarios:

- Can the multiple teams collaborate so they work on the same feature set? You can, assuming you have tools and communication organized so the teams share a virtual workspace.
- Reconsider the feature sets so you can create minimums. How can teams define and complete work on a Minimum Marketable Feature or a Minimum Viable Product? Once the team finishes

these minimums, this team can move to other work (new feature set, technical debt, or defects) and another team can take more work on this feature set later. (See *Create Your Successful Agile Project,* ROT17 for more details.)

- If you need more teams to work on a given product release, maybe you have a program. Consider how various leaders in the organization collaborate to shepherd the business value of the product across the organization and resolve impediments across the feature teams. Consider *Agile and Lean Program Management,* ROT16A.

Remember that "scaling" is often not the right metaphor for what you need. You might need to reorganize the work or the teams to accomplish the work better.

10.7 Integrating People Traps

When we think about hiring or changing people for a distributed team, we've seen these traps:

- Visitors help a team finish work.
- We don't have to change our sourcing.
- We can save time by delegating up to managers.
- Digital nomads can't work for us.
- We can hire an expert anywhere and make the team work.

10.7.1 Trap: Visitors Help a Team Finish Work

Not all agile teams have all the capabilities and skills they need to finish features and release. Too often, managers want to add "visitors" to a team to help them do so.

We don't recommend this approach. Both collocated and distributed agile teams have trouble integrating visitors into their teams because those people have not participated in building respect and empathy with each other via working agreements and delivering work.

Some specific challenges occur for the teams and the visitor. If the visitor is asked to assist "part time," no one knows when the visitor is available for the new team or the visitor's home team.

Sometimes, visitors influence the team's estimates. If so, the team may need to rethink their assumptions and actual estimates. The team may even need to re-estimate if the team required the visitor's expertise.

If your team does require some temporary expertise, consider these options:

- Walk the visitor through the working agreements, so this person understands.
- Mob on the work, so the team learns what the visitor knows.
- At minimum, ask the visitor to pair with one person on the team.
- Either have the visitor on the team or not at all.
- If the team does not feel they can utilize the visitor's skills in a full-time capacity, treat them as a consultant who does not do the work, but advises about the work. Then, ask the visitor to teach team members the skills they need.
- Manage the risk of the visitor by clarifying when the visitor is available and the duration of the visit. The clarification may help avoid estimation and delivery problems for work that depends on the visitor after their departure.

Too often, managers believe this trap because they believe in resource efficiency. See *Think in Flow Efficiency* on page 10.

10.7.2 Trap: We Don't Have to Change Our Sourcing

Depending on your market and products, you may need to reflect on and change where you source new candidates.

You may have been successful in one geographic area using one form of sourcing. You might need different sourcing mechanisms in a different parts of the world. Referral programs help sourcing. However, you might need to find candidates faster than your referral program discovers them.

Define your overall sourcing strategy so you can discover diverse people: gender, personality, and experience at a minimum. Also consider the potential cultural fit you seek. Then re-examine where you might encounter people who could fit your hiring profile.

- Would they be on websites where you advertise?
- Would they be on websites where your current teams read reviews and seek advice?
- At what types of conferences would you find potential candidates?
- Can you also find some online events to speak about "how you work"?

Consider alternative sourcing strategies for maximum benefit to your hiring process. Re-examine these strategies at least twice a year to see if they still work for you.

10.7.3 *Trap: We Can Save Time by Delegating Up to Managers*

Too often, managers think they can source and interview candidates, decide, and "integrate" new hires on behalf of the team. Managers think they will save the team time by having the team delegate "up" to the manager.

The problem is that the team has to make the hiring decisions, and the team has to integrate the new team members. The manager is often not aware of the nuances of the day-to-day work of the team. Team members focus on these nuances in the interviews.

If your team feels short on time for finding and integrating candidates, consider these options:

- Move to continuous hiring, so the team always sets aside time for hiring activities. If you're always hiring, no one person spends more than an hour or so a week on hiring activities.

- Rotate a subset of the team through the interview team with a hiring matrix.
- Rotate developers from similar teams onto the interview team.
- Mob wherever possible: on sourcing, interviewing, and especially integrating a new team member into a team.

When team members find, hire, and integrate new people onto their team, they create collaboration opportunities and make fewer hiring mistakes.

10.7.4 *Trap: Digital Nomads Can't Work for Us*

Some people like to change location to follow family members, follow warmth or skiing, or just because they like to move around the world. Some people choose different venues based on the type of work they plan to do for the day. Being dispersed actually makes this easier. However, teams need working agreements to make the work easy for the digital nomad and for the rest of the team.

Once a team has integrated a nomad, it doesn't matter where the nomad works. As long as the nomad has sufficient technical infrastructure, they can schedule when they work and make and meet commitments to the team.

If your organization requires "policies" for digital nomads, consider security policies rather than any other policy. If a team creates a collaborative environment that delivers value, do you care where anyone is?

One distributed team chose to adjust for the digital nomad. For the first month working with this small team, the nomad made very valuable contributions to the work. Then he asked the team lead if he could possibly work from another location as long as he made the same contributions. They agreed. So the nomad moved to Thailand. He adjusted his work hours to make daily standups and other meetings. Once the team realized his "work hours" may not work well for his "sleep hours," the team offered to adjust some of their meetings.

We have seen team members be a little reluctant to share their change of location with the organization, because they work so well with their team. However we encourage transparency with location, so the IT organization can verify the security settings for the nomad. Access through VPN helps solve this problem.

10.7.5 Trap: We Can Hire Experts Anywhere

One manager said, "There are smart people all over the world. I want to hire anyone, anywhere, at any time."

Maybe.

When managers hire experts "anywhere" they create isolated-by-design people.

Ann, a CIO in Ohio, under pressure to deliver several projects, contracted with an offshore team to provide testing for the projects. All the developers and the Product Owner were on several floors of one office building in Ohio, and the tester was in Hyderabad.

She chose the Indian provider because she'd worked with them before, with great success. She expected the same this time.

The contract testers were capable of doing the work. And, because of the insufficient hours of overlap, the teams never succeeded in delivering on a regular cadence or finishing the projects fast enough.

The contractor slowed the project, because they were isolated-by-design.

We've seen the same with employees. It's not about the relationship of the isolated person to the company. It's about the relationship—or lack thereof—with the people on the agile team. Even if your motives are good, such as funneling work to people who need it, agile teams need the personal relationships.

One company, based in New York, with founders from China and Russia, wanted to funnel money back to their colleagues in their home countries. They had isolated-by-design experts—people very far away from their colleagues on the supposedly agile teams. The teams had trouble finishing anything in a reasonable time. The execs saw no need to change because the arrangement met their goals. However, the teams had trouble being teams and they suffered along with the value they tried to deliver.

Make sure you don't create team problems with people who are isolated-by-design. See *Default to Collaborative Work* on page 36.

If you do know of someone with few hours of overlap and you want that person, ask yourself these questions:

- Is it reasonable to hire a complete feature team where that expert is?
- Is it reasonable to ask this person to spend significant time with the rest of the team, working in a collocated fashion for a few weeks, so they can learn from each other?
- Is it reasonable to ask the expert to alternate time-shifting with the rest of the team?

You often have more options than you might consider when you think you want to hire someone many hours away from the rest of your team.

You can hire feature teams all over the world and they can succeed. Hiring individuals all over the world without sufficient hours of overlap makes team members crazy and delays the product development. See *Avoid Chaos with Insufficient Hours of Overlap* on page 41 for more details.

10.8 Now Try This

We've only highlighted hiring issues specifically for distributed agile teams in this chapter. You may have other considerations for your team and culture.

1. Consider the interpersonal skills you need for your team.
2. Consider how you will "test" a candidate's ability to manage hours of overlap, the candidate's work in progress, and how the candidate will work collaboratively with the rest of the team.

3. Consider how the team will organize their hiring process to best manage their time and still understand how to assess a candidate.

Now, it's time to address the role of the executives and managers for a distributed agile team.

CHAPTER 11

Lead Your Distributed Agile Teams to Success

We started this book discussing the three mindset shifts required for successful distributed agile teams: manage for change; emphasize communication and collaboration; and use agile principles in *Distributed Agile Teams Are Here to Stay* on page 1. We also discussed how to *Think in Flow Efficiency* on page 10 to support a team's frequent delivery of value.

Throughout the book, we have discussed the eight principles to support the three mindshifts in *Focus on Principles to Support Your Distributed Agile Teams* on page 23:

1. Establish acceptable hours of overlap.
2. Create transparency at all levels.
3. Create a culture of continuous improvement with experiments.
4. Practice pervasive communication at all levels.
5. Assume good intention.
6. Create a project rhythm.
7. Create a culture of resilience.
8. Default to collaborative work.

Leaders who live these organizational principles are able to coach, serve, and create a successful distributed agile environment. That environment allows the entire organization to succeed.

How can you serve your teams?

You'll notice we used the plural of team—teams. That's because we don't know of any organization that has only one distributed agile team. Organizations we work with have multiple agile teams. The exception is one-team startups and they quickly grow into multiple teams when they are successful.

Teams don't succeed on their own with these principles. The decisions managers and executives make affect how well these teams can succeed.

Let's start with how a manager's actions are amplified—and how you can use that to create successful distributed agile teams.

11.1 Cultivate Affinity Between People and Teams

Team affinity is not sufficient for a distributed organization. Successful distributed agile managers cultivate affinity between teams and managers as a key management practice.

How can you help teams discuss issues at the team level? How can you ensure that the right communications pervade the entire organization? How can you monitor delays or misunderstandings?

"Cultivate affinity" signifies that you are never really done with encouraging affinity within and between teams. As a leader in your organization, consider how you can monitor, encourage, and connect people to grow the affinity in your organization—and possibly with your customers and suppliers. If not, you will quickly build up other forms of organizational debt that will slow you down. (See *Set a New Direction* on page 240, for more information.)

As a leader, you can amplify or destroy distributed agile teamwork. *See Your Team Type Traps* on page 82 for problems we encounter most often.

Let's discuss actions you can take to create an environment that works for your teams.

11.2 Create an Environment to Amplify Distributed Agile Teamwork

Distributed agile teams need organization support in the form of tools and culture to answer these questions:

- How can we form distributed agile teams with acceptable hours of overlap to leverage worldwide expertise, support collaboration, and enhance value delivery?
- How can we create sufficient transparency so any distributed team member not only understands what they are working on but how it helps the business and customer?
- How do we encourage continuous innovation through experimentation so teams are always improving how they deliver and generate new market-disrupting ideas?
- How can we keep teams and team members aligned by encouraging open communications of problems and successes, weaknesses and opportunities, and how to best leverage the strengths of the different individuals and teams?

As an executive or leader, think about how your answers to these questions. Might you change the supporting environment for your teams through experimentation? Such exploratory changes can amplify the agile culture and create a successful distributed agile environment.

If you can, invite the teams to collaborate with you on creating an environment in which they can help answer these questions. In an agile organization, managers don't need to have all the answers.

What if you can't create teams with acceptable hours of overlap? Or, you can't create sufficient transparency? Or create experiments? Or any of the other necessary conditions for an agile team to succeed? See *When Agile Approaches Are Not Right For You* on page 17.

11.3 How Leaders Can Show Their Agile Mindset

We've seen many executives and managers feel substantial pressure to "deliver." Someone, not necessarily the people currently in management positions, decided they would divide-and-conquer the work to make the best use of people and reduce project cost. We discussed these fallacies in *Think in Flow Efficiency* on page 10.

You might want to collect data for your organization about project cost, and cost of delay due to resource-efficiency thinking. Start with measuring the Cost of Delay in your distributed teams, and consider measuring project costs.

11.3.1 *Measure Costs of Delay in Distributed Teams*

The Cost of Delay is one or more of the following costs due to delaying a feature or a project:

- The company misses the potential sales from the introduction delay.
- The delay introduces lower maximum sales.
- The delayed introduction creates less overall demand, so the feature has less value in total.
- With less demand, the end of life might be earlier with a delay. Often, this is a function of not being able to capture the market at the optimum time. Since you don't have the customers, you have an earlier end of life.

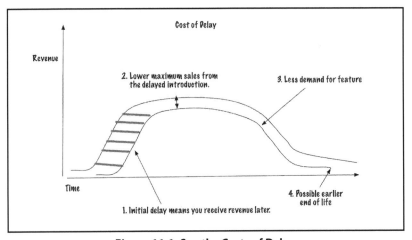

Figure 11.1: See the Costs of Delay

You can measure a simplified cost of delay if you measure the team's cycle time, as we suggested in Map the Value Stream to

Visualize Cycle Time. First, take one or two "average" features or stories. Measure the wait times until the entire team finishes the story. Then, measure the total cycle time.

How much of the total cycle time does the team spend waiting?

The more time the team spends waiting, the more likely the team doesn't have sufficient hours of overlap. Maybe, the team doesn't have sufficient communication tools so they can't easily collaborate. Maybe, they're waiting for someone else in another part of the organization. (See *Diving for Hidden Treasures*, RE14 to review other possible costs of delay.)

The team is in the middle of this "mess." They might not realize they have impediments because that's just how it is for them. You, as an organizational leader, can help the team visualize their wait states. You might even help them identify problems and facilitate alternatives to how they can manage their costs of delay.

11.3.2 Measure Project Costs

Throughout this book, we've assumed the teams have people assigned 100% to one team. No one multitasks on several teams. No one is a visitor on other teams.

When teams take work as a team—agile approach or not—it's trivial to measure project cost. Each team has a run rate, based on the people on each team and their salary plus overhead rates. Add each person's run rate, multiply by five for the number of days in the week, and now you have the team's weekly run rate.

You can calculate longer run rates by multiplying the number of weeks with the weekly run rate. If you want to estimate how much more, use gross estimation to estimate the rest of the project duration in weeks and then multiply by your team's run rate.

If you do not have the ability to assign people to teams 100% of the time, you will have a much more difficult time calculating project cost. That's because you don't know when a person is available to work on your team. In that case, we recommend using the team's median or average cycle time. Use your judgment as to which one.

We like to consider the worst case scenario: People are rarely available and the cycle time is long. The weekly run rate might decrease. However, the entire project duration increases and the project cost increases.

In that case, take the maximum (or median) cycle time, the maximum run rate for the team, and use that as your measurement. Highly dispersed teams who do not have sufficient hours of overlap might not cost much for a given week. However, the projects often take many more weeks or months of time to complete.

To create the least-expensive projects, work on reducing project duration *and* the team's cycle time:

- Keep teams together so they can depend on each other for their learning and collaboration. That helps the team reduce delays due to knowledge gaps, increase their throughput, and decrease cycle time.
- Locate (if you can) all team members within at least four hours of overlap to reduce communication and collaboration delays, so they have the shortest possible cycle time.
- Eliminate off-team multitasking, so the team can create the smallest possible cycle time.

These three points help a team decrease cycle time for a given feature. And, they help a team decrease overall project duration.

11.3.3 *Shift Your Management Mindset*

Thinking about projects, teams, and people this way might be foreign to you. When you measure costs of delay and cycle time, you find opportunities to change the culture of the organization.

Instead of thinking about project costs, consider what you might invest to reduce cycle time or cost of delay. How fast would those investments or changes pay off and reduce cycle time and therefore project costs?

You might not be able to make these changes on your own.

When you make the cycle time and cost of delay visible to your team, the team can help generate ideas on how to reduce those impacts. You need your team to make these changes. You do not control the changes alone.

When you make the cycle time and cost of delay visible to your peers or managers, you might be able to change the organization.

You don't have to control people in an agile organization. Offer teams and management the transparency to visualize key indicators, such as cycle time and cost of delay. When teams see the effects of how they work, they might change their practices. When managers see the effect of their previous decisions, they might change their decisions.

Creating transparency and pervasive communication of this data might create a collaborative culture shift in your organization. Consider how you might build or rebuild your agile management skills to create this culture shift.

11.4 Build Your Distributed Agile Management Skills

You might already realize that every leader's action gets amplified when other people see it. What you think is an off-hand comment might change the course of a given team's direction. Distance increases that amplification.

You can amplify things that help your organization, such as transparency, continuous improvement, and communication. But you must also monitor for challenges that can become amplified in a distributed environment such as misunderstandings, miscommunications, delays in coordination, and unequal notification of key changes to the business process or direction.

Managing a distributed agile organization challenges the best managers. Consider how the principles express themselves in agile management skills and where you might want to experiment.

Principle	Management skills that reflect this principle
Acceptable hours of overlap	Support the team in selecting their optimal collaboration times, and knowing when they can't collaborate using an agile approach.
Transparency at all levels	Support transparency in the team, across teams, across the organization.
Culture of continuous improvement	Encourage frequent experimentation, measurement, and learning.
Pervasive communication	Treat every employee as a business partner.
Assume good intention	Practice empathy, not blame.
Create a project rhythm	Encourage your team's rhythm by creating a cadence for your work.
Culture of resilience	Use resilience as a form of risk management.
Default to collaborative work	Work with others as part of effective teams.

Figure 11.2: Management Skills Based on Distributed Agile Principles

Let's discuss each of these skills and how managers can practice them.

11.4.1 Model Acceptable Hours of Overlap for Your Teams

Your predecessors may have made decisions to invite people all over the world to be on one team. It's possible you hired people by acquisition. Regardless of the causes, you now have teams with people all over the world, without acceptable overlap.

Consider how you can support the team to create their acceptable hours of overlap:

- Support flexible work hours
- Support people with the necessary collaboration tools and infrastructure for their team
- Adjust team composition, with team input, to take advantage of closer hours of overlap

If you invite the team to find their acceptable hours of overlap, they might be able to do so without any management help. Teams are great at solving problems if they see a reason to solve the problem.

If you model the behavior of reasonable hours of overlap in your dealings with the distributed team, you can show them how they might approach the problem. Explain that you are not willing to attend a meeting at 3:00 a.m. your time and you don't expect anyone else to do the same. Look for opportunities to shift your time to closely collaborate with a team to set a vision or clarify a new initiative.

In turn, teams might decide to share the pain and adjust their hours to get minimum hours of overlap for collaboration. They might decide to use a backchannel, a copilot, or to use any of the other approaches we mentioned to combine collaboration forms and channels.

Avoid telling the team members they *must* change their hours. People can choose to change their hours for their good and the good of the team. They are often not willing to go along with a management "mandate."

They might decide that the team they have is not the team they need. That's where you might be able to assist them in creating a team that works better with acceptable hours of overlap.

11.4.2 *Model Transparency*

At the very least, consider sharing your impediment removal work for your teams. What else do you feel comfortable sharing with your teams? Your other work? The latest sales wins against competitors? New awards won by the company? New opportunities?

What do you feel uncomfortable sharing with your teams? Losses to competitors? Why there was a loss to competitors? Potential pivots in the company direction? Possible downturn in the market overall? Hiring freeze or salary information?

Experiment with sufficient transparency so your team knows what they need to know and when:

- Encourage questions out in the open
- Support open workspaces so teams can easily learn from each other
- Create equal access to all tools for all people on the team

The more you as a leader can share your uncomfortable topics and why they are uncomfortable for you and the business, the more the teams will understand why it is critical to deliver sufficient value in a timely manner. If teams do not understand the competitive landscape, how can they build features that will survive the market? If you do not show teams how competitors position their products and how your product doesn't build on its position, how can your teams successfully respond?

If you can't share hiring freeze information, people don't understand how necessary their contributions are to improving the revenue for the organization. You might not be able to share salary information, but do consider sharing the salary levels and expectations for a given level.

Transparency is about sharing the good with the bad and understanding what you all can learn from it. If you cannot share this level of transparency with your teams, they are unlikely to be transparent with you.

11.4.3 *Model Continuous Improvement*

Successful agile teams expect to improve on a regular cadence so they can improve their team's process and product. How can you as a manager show how you improve?

Consider how you can create continuous improvement opportunities for your teams:

- Experiment as a manager and with other managers and staff to create the best possible distributed agile environment for your teams.

- Encourage experimentation in the team and across teams.
- Offer opportunities for in-person learning as often as makes sense and at least once a year.

Managers can use and model double-loop learning, also. (See *Cultivate Continuous Improvement with Experiments* on page 26 for more information.)

 Demonstrate your own in-person learning.

You might want to be transparent about the obstacles you're working on and removing for teams. You might want to show how you change your management or your reports or something you do, so the teams can see you are modeling continuous improvement.

People don't often expect managers to proceed with their work at the same progress as the teams do for features. People recognize that managers tend to work on human systems, which are more difficult to change and require more people than one team. And, if you show your improvements and learning over time, people will respect you more.

Collaborative experiments work best when you understand your collaborators. The more distributed you are and the more you want to take advantage of agile approaches, the more important periodic in-person events are for the people on the teams. Consider how you can bring your entire organization and smaller groups within your organization together to form broader experiments beyond a single team.

11.4.4 Model Pervasive Communication

Even in collocated agile teams, managers need to say things more than once. People hear and interpret meaning in various ways. So, we suggest you err on the side of too much communication, rather than too little. This may include communication repeated over various channels, such as a real-time meeting, email, or internal blogs.

Consider how you can treat each employee as a business partner:

- What could you share so your employees understand how the business is doing, e.g., profits, company sales, company concerns, how the board measures success, and maybe more?
- What do you need to share in terms of your personal learnings so people see you as human and as a model for their distributed agile approach?
- How can you clarify project and organizational goals?

When do you know you have communicated enough? When you start to hear a similar message coming back to you from your teams, you know they received the message. When they tell you it's "too much," that may be enough communication on one topic also.

One set of teams had grown through hiring. They had a very competent agile manager, Sue, who the teams liked. The teams continued to hire and split, hire and split. They grew to four large teams of at least nine people each. Sue explained to her manager she needed help. The leadership team agreed.

Sue helped hire Frank, an agile manager with similar qualities to Sue. Frank set up one-on-ones as part of his integration into the company. As part of his one-on-ones, he asked the team members if they had questions. The most repeated question Frank heard was: "What's happening to Sue? Is Sue leaving?" The teams really liked Sue and did not realize how overloaded Sue was.

Even two weeks later, people were still asking where Sue was going, even though Frank explained in multiple meetings that Sue was staying and shifting responsibilities.

Sometimes you must repeat the same message several times and several ways.

11.4.5 Model Understanding to Assume Good Intention

If an interaction online puzzles you, and you pick up the phone or get on a video call to clarify your concerns, you model this skill. It's easier to walk down the hall if you have a collocated team.

Especially as a manager, people take their cues from you. If you are willing to extend courtesy and assume the best of people, they may be willing to do the same. See *Break the Email Chain* on page 182 for an example.

Consider how you can help your teams treat problems as learning opportunities for the team and not a blame game:

- Use phrasing such as, "What would have to be true for this person," in your communications so people realize you are practicing empathy.
- Ask for help so you can understand the issues challenging your team.
- Offer to facilitate synchronous conversations (with video) when asynchronous conversations show misunderstandings.

It's very easy to assign blame to people at distance. Instead, practice empathy. You will realize the team has impediments they can't see, and you'll serve the teams so they can be more effective.

11.4.6 *Model Your Project Rhythm*

Managers rarely use timeboxes in the same way as a team. However, managers can provide a cadence of one-on-ones, learning, explaining corporate information, and more. The more you can build and maintain a cadence for your work and the people you serve, the more they know what they can expect, and when.

Consider setting these kinds of cadence expectations: when the team members will have one-on-ones with you, when the team might have formal team learning, and when the team will get together in person.

Consider how you might create a cadence with your team:

- When you have one-on-ones with the people
- When they provide demos to themselves, to you, and the greater organization
- When you report back to the people and the team about removing impediments

When you model your rhythm, you can help the team model theirs.

11.4.7 *Model Your Resilience*

Too often, we see managers who want distributed teams to "stick to the plan," when it's clear the plan no longer works. Keep in mind that *Work with Humans Requires Empathy* on page 163.

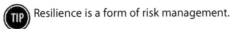 Resilience is a form of risk management.

When teams are resilient, they can experiment with the product and the process. They can create and consider multiple options. The team members can help each other, in new and different ways. But the team cannot change product and process without considering business outcomes.

When managers are resilient, they focus on clarifying the business outcomes and serving the people to achieve those outcomes. The agile manager becomes a servant leader to the distributed teams and the business.

Consider how you can serve your teams to create and refine their resilience:

- Support socialization. When people don't have a sense of affinity, they are faster to blame. When people know and understand each other, their system of work and their product context, they create empathy and affinity with each other. When managers provide an environment of socialization, the team members can learn to forgive any one team member's "temporary bout of insanity," as one manager explained. Without some form of affinity between people, the people can't build trust or empathy. That breaks the idea of a team built on affinity.
- Support shared goal setting so distributed team members build deeper connections. When people share their professional and personal goals, the team members can create opportunities to

collaborate or, at some point, transfer that work to the person who wants to learn it. This is another form of affinity, where team members help each other with their goals.

- Some agile teams don't see their bottlenecks or bottlenecks between teams—work stuck in one place or with one person. You, as a leader, might help your team learn to see their bottlenecks and create options for flowing work through the team.

- When team members share enough about their home context, the team understands anything that could prevent the team members from full participation at varying times. Many people find this kind of sharing quite difficult. Who wants to admit that their two-year-old is having the 67th ear infection this year? Or even that they need to take time to manage elder care. People are human and they have human needs. We are not suggesting that you run the business as a charity. The more people know about each other, the easier it is for the team to create resilience and quickly support each other.

You can model your resilience when you ask for and consider alternatives for everything: for plans, for tools, and how you and the team uses them, and for how the team accomplishes the work.

You might even consider asking for more resilient approaches as you model your continuous improvement, so you can create experiments.

The more resilience you create and practice in your management work, the easier it is for your team to do the same in their work.

11.4.8 Model Collaborative Work

Too often, managers think they need to work alone. They don't ask their team or colleagues for help.

Effective teams build psychological safety so they can offer and receive feedback on their work. The more the people collaborate, the more effective they are as a distributed agile team. Collaboration allows them to understand the problem space and the possible options available to them. Collaboration allows them to learn together instead of keeping knowledge in individual silos. Collaboration allows them to explore risks together and choose the options that deliver value quickly and with high quality.

Distributed agile managers have exactly the same needs. The more they work with the teams and across the organization as management teams, the easier it is for the manager to be effective.

Management collaboration—with other managers and with the teams—helps everyone see how to collaborate across distance. For example, you might collaborate with product management people on the product direction. You might work with the project portfolio team to help them assess the relative ranking of each product. You might work as part of a customer task force to understand customer needs more and how those needs affect the product.

And, you might work with your teams, to co-plan the business outcomes so the team can manage its own work.

Assess the work you do alone now. Consider how you can serve everyone's learning—including your own—if you can work as part of effective teams.

11.5 Set a New Direction

We meet many managers who inherit challenges to distributed agile teams. Their organization hired people around the world with the expectation of lower costs. Or, they grew by acquisition and the acquired teams were not provided an opportunity to create affinity with the rest of the organization. Or, people moved many hours away and the teams don't have sufficient hours of overlap.

Whatever the reasons, these managers now realize they have organizational debt. The debt is in the people-organization—the people aren't organized to fit the business and their human needs.

Here are some quantitative ways to see if you have organizational debt. First, map the value stream. Now, ask questions:

- Are people waiting for the right environment, such as licenses or other access to tools? If so, use money to fix this problem.
- Are people waiting for the right capabilities and skills to deliver value? Do enough people have sufficient problem-space domain expertise, such as in the Product Owner role? Do enough people have solution-space domain expertise so they can design the code and tests and coach others into more knowledge? If the team doesn't have all the skills they need, they will too often blame each other or management or customers instead of working together to solve the problems.
- Are people waiting for technical answers? Learn which answers they are waiting for. If it's code review or story ranking, ask the team to explore how to plan together, what their normal collaboration modes are, and if they think they have sufficient hours of overlap.
- Are people waiting for approvals? If team members need permission for deployment or other activities that increase cycle time, the problem is most likely an organization-wide, systemic problem. You might need to solve that at the management level.

As you review the value stream map and answer these questions, you might realize this "team" is only a team in name. They will continue to wait unless you or the team changes who is on the team. Consider looking at the qualitative reasons before you change the team members.

Here are some qualitative reasons your team might be a victim of organizational debt:

- The team doesn't know how to work as a team because they haven't been trained in agile approaches. Or, they don't know how to work together. You can address this problem by bringing the entire team together for two weeks: the first week is for agile training and practice as a team. The second week is learning how to work together.

- The team members don't have psychological safety. They have not practiced their deep interpersonal skills. Consider offering training about how to use and practice those interpersonal skills.
- Listen for people who assign blame. People who blame are often afraid they will get blamed for something "wrong." Too often, the culture reinforces that blaming stance. Ask yourself how you react to bad news. Do you blame people for missing a deadline or other "failures?" If so, you are reinforcing a culture of blame instead of a culture of learning.

If you've considered the quantitative and qualitative options, and your team is still not as effective as you think they could be, consider when to change the team membership.

We normally recommend that the team decide how to organize itself. If you are not quite ready for that, decide how to create a cross-functional team with sufficient hours of overlap as a first step. Then, apply the other recommendations throughout this book.

Whatever you do, don't assume things will "just get better." The team cannot improve its situation alone, especially if they've been working under organizational debt for any length of time.

All of these approaches require listening to people. You might have to travel to where the people are. You might have to facilitate a one-week or longer meeting so the people can connect with each other.

A manager's best tool is still a one-on-one conversation. The second-best tool is management by walking around and listening. If you're not listening to the people in some way, are you fulfilling the promise of management for those people?

We are not fond of distributed agile managers always traveling from one place to another. However, if you are a manager or executive, you owe it to the people in various locations to get together so you can hear them.

This book has focused on applying the agile practices within teams. Managers of multiple teams also need to pay special attention to how these principles are applied across teams.

11.6 Focus on "Better" When Scaling Distributed Agile Teams

The agile community talks a lot about "scaling." We discussed how to split teams to scale to more teams in *Split to Scale Your Teams* on page 215 to maintain the agile culture when hiring new people. In addition, scaling might mean sharing agile approaches from one team to another or one product to another, as in a corporate agile transformation. The third scaling meaning might be how to link the efforts of multiple teams so the company succeeds with delivering a larger product.

Many people focus on "bigger"—more process. Instead, how can you amplify the "better" when scaling your agile efforts?

We've seen teams get "better" for their scaling efforts when they focus on experimentation, their collaboration, and using the agile principles to create environments that allow the teams to succeed.

We have seen teams stall because of the process when they attempt to apply frameworks across dispersed people and teams. Just as we recommended you use principles to adapt any given team's practices, we also recommend you consider principles to scale the efforts across your teams.

Consider the principles as you try to scale or coordinate the work of multiple distributed agile teams:

- What does acceptable hours of overlap mean?
- What is transparency across distributed teams?
- What is continuous improvement across distributed teams?
- What does pervasive communication look like across a program of distributed teams?
- How can teams practice assuming good intentions on the part of other teams?
- What does program (a collection of teams or projects that collaborate to deliver one business outcome) rhythm look like?
- What does resilience mean across these teams?
- How do multiple teams default to collaborative work?

There is no one right answer for all of these questions. If you have programs, consider the ideas in *Agile and Lean Program Management*, ROT16A.

If you're trying to use agile approaches to create business agility, consider how well your organizational culture supports agile approaches and the principles in this book. Business agility is outside the scope of *this* book.

11.6.1 Options for Acceptable Hours of Overlap Across Teams

If more than one team is working on a collection of related features in the same area of the code base, there may be a few options for overlapping the hours of multiple distributed teams to support collaboration:

- You might set up a meeting where representatives of each team can raise issues and concerns on areas of collaboration.
- Pull members from each team to form a new team for a *brief* period of time to collaborate on any interfaces that will be used across the team. Once the interface is complete, team members return to their original team.
- Consider asking two or more teams to work as a larger team for a brief period so they can learn each other's domains and possibly create different team makeup.

The more each team has acceptable hours of overlap *as a team*, the easier it is for them to decide how to work with other teams. Minimize the need to collaborate across several teams. If each team has to figure out how to work as a team and how to work across teams, they may never be able to collaborate.

11.6.2 Options for Transparency Across Distributed Agile Teams

The more teams on a given effort, the more difficult the communication becomes between teams and team members. People find it difficult to stay current with what everyone else is doing.

When teams need to work together, they need to know where each other's information lives, and how to access that information. Our empirical evidence (not backed by research), is that three to five teams is the tipping point.

Consider asking whether the teams need to collaborate on the same code and test bases, or if they need to see each other's demos. If teams need to collaborate on the same code and tests, it's easy for team members to step on each other's work. If the teams need to see finished features, they might be able to be transparent about the state of each feature.

- Allow and encourage teams to share visibility of their boards with each other. You don't need a tool for this. Images in a known place on your intranet work just as well as tools. Possibly better.
- Have a combined board that shows the interdependent work between the teams. Work unique to the team stays on the specific team's board.
- Have combined planning and review sessions for the teams that need to collaborate. If you realize that more than three teams need to collaborate, check to see if you have siloed teams, instead of cross-functional teams.

11.6.3 *Options for Continuous Improvement Across Distributed Agile Teams*

Many distributed agile teams realize their challenges for improvement intersect with other teams.

- Conduct multi-team retrospectives in-person. These type of retrospectives can be valuable when a feature set or large initiative is delivered.
- Conduct multi-team retrospectives online. Online retrospectives can be conducted online for feature sets in progress as teams deliver increments.

- Conduct hackathons that encourage people across different teams to collaborate on possible innovations.

Teams see the need to fix problems in the process and the product. Encourage and allow the teams time to do so.

11.6.4 *Options for Pervasive Communication Across Distributed Agile Teams*

Multi-team programs need to balance important communication between the teams while isolating communication that is unique to one team. Otherwise, the issues overwhelm all the teams. People can miss important information or changes.

- Have a cross-team meeting once a week where anyone can bring up key updates, dependencies, and blockers. These messages will likely be echoed in other channels.
- Have a cross-team backchannel where any team can easily reach out to any other team and offer visibility to all other teams.
- Create communities of practice for intentional learning across teams.

When multiple teams work on one effort, they can miss key information. Create several channels so people can hear and see this information.

11.6.5 *Options for Assuming Good Intentions Across Distributed Agile Teams*

When people see each other's work, they can understand the context. When people can easily talk with each other, they can treat each other with respect. When people are on the same team, seeing each other and each other's work becomes possible. When people are on different teams, they might have a more difficult time seeing each other's work.

Leaders who focus on transparency and a belief in good intention can help the teams create these possibilities.

Congruence applies not just to people, but to teams. See *Assume Good Intention* on page 29.

If you want multiple distributed agile teams to work in concert, find ways to:

- Help each team finish something every single day and check it into the code base and build the product.
- Help each team be transparent in their work.
- Reduce feature interdependencies.

Too many teams don't assume the best of each other because the work isn't transparent. When teams can't create an entire feature through the architecture, they fear the interdependencies that prevent any of the teams from making "their" deliverables.

When managers think in flow efficiency and think about what's good for the *product*, the managers can help the teams assume good intent. For more information, see *Agile and Lean Program Management*, ROT16A.

11.6.6 *Options for Program Rhythm Across Distributed Agile Teams*

In our experience, larger efforts mean more change. Assess your need for change or release across the teams. Do the product owners need to change backlogs often? Do your customers need more frequent releases?

Here are some considerations:

- If you're using an agile approach so each team can see its progress, but you don't need to change the backlogs or show the customer progress, consider a quarterly program rhythm. The more volatile the market, the more likely you need to make a more frequent rhythm.
- Consider using one- or two-month rolling wave plans to refine the backlogs, demonstrate, and retrospect, even if you do use quarterly planning.

- Consider moving to a more continual planning approach. If you have feature teams who are able to work independently, this option offers the most capability for change, demos, and releases.

Assess your needs to for creating plans, seeing demos, and releasing products. Then, you'll be able to create a rhythm that works for your program.

11.6.7 Options for Resilience Across Distributed Agile Teams

If you have just one team with specific expertise, what happens if they all want to go on vacation at the same time? Or, if you have small teams, how can other teams help if a team needs help? That's the idea of resilience across teams.

- Ask the teams when they want to collaborate with other teams to build domain expertise in each others' areas.
- Create communities of practice with intentional learning every week or two.
- Bring teams together in person every so often, to work together to learn from each other. That prevents siloed expertise on teams.

Watch for expertise bottlenecks in the teams' backlogs. If one team has a year or more of work queued, that team has siloed expertise. This might be expertise worth spreading to other teams.

11.6.8 Options for How Teams Might Default to Collaborative Work

We don't often think about how multiple teams might collaborate. If you do have work that spans multiple teams, consider what collaboration might look like.

- Ask teams to demonstrate together, so everyone can see how the product works.

- Consider asking the teams to collaborate on standards. For example, the API, or a look-and-feel for the product.
- Consider how multiple teams might focus their efforts on a prototype, research work, even a particular feature set to make progress.
- Multiple teams might not need to collaborate. If they do, ask them how they want to collaborate.

11.7 Start with a Distributed Agile Management Culture

We've suggested ways to think about the team—how the supporting environment works, and how that environment helps the team work better. We've also suggested ways to do the same across multiple teams.

We often discover that successful distributed agile teams work because the managers create an agile culture. Often, that's because successful managers start with their own thinking and the culture they want to create and reinforce.

Use the principles we've described throughout this book to create a distributed agile management culture. Consider your possible actions:

- Create management teams with sufficient hours of overlap. These management teams can then make timely decisions for backlogs at all levels: the project portfolio which implements the corporate strategy, and the product backlog which helps teams know the most valuable features and experiments.
- Build and maintain single-focus feature teams with acceptable hours of overlap. Single focus means everyone on the team has the same goal, everyone is fully committed to that one team, and the team has acceptable hours of overlap.
- Create an environment in which team members and teams who work together on a regular basis see each other face-to-face at least several times throughout the year.

- Allocate sufficient money for every team to have the virtual teamspace, including sufficient communication tools.

Consider how your management practices serve your distributed teams, agile or not. Then, consider how you can use these eight principles to create a better environment for your teams.

11.8 Set the Path for Your Distributed Agile Journey

Distributed agile teams can work, and work well.

Successful distributed agile teams live the principles in a supportive culture. As evidence, some successful distributed agile organizations have employee retention rates well over 90 percent and lead their markets—good success measures.

As leaders in the organization, you may need to change the culture, the makeup of the team, and possibly your perspectives on how people work and how to measure their work.

This isn't easy. It's not fast. It's worth it. Your organization will be more resilient. Your teams will be more effective. The people will be more satisfied. Your customers will thank you.

Throughout this book, we've suggested guideposts through the principles. Use the ideas in this book to create your path to your successful distributed agile teams.

11.9 Now Try This

Ask yourself these questions:

1. How do you currently model the eight principles we have outlined in this book? You may want to refer back to *Distributed Agile Teams Are Here to Stay* on page 1 and *Focus on Principles to Support Your Distributed Agile Teams* on page 23.

2. How often are you surprised by the your distributed teams now? Are the surprises positive or disturbing? Did you discover

the surprise was an amplification of something you said or did as a leader? If the surprises are disturbing, see *Avoid Chaos with Insufficient Hours of Overlap* on page 41.

3. Where do you see communication challenges between your teams? Refer back to *Identify Your Distributed Agile Team Type* on page 65, *Communicate to Collaborate* on page 87, and *Create Your Collaborative Team Workspace* on page 111.

4. How can you help your teams in setting the right culture to deliver value on their specific products while working well with other teams within your organizational culture? See *Cultivate Your Distributed Team's Agile Culture* on page 129.

5. How can an agile approach support your evolution of team culture? See *Build Respect with Working Agreements* on page 147, *Adapt Practices for Distributed Agile Teams* on page 173, and *Integrate New People Into Your Distributed Agile Team* on page 197.

As you answer these questions, you'll see how you can lead your distributed agile teams to success.

We wish you well on your journey. Let us know how it goes.

APPENDIX A

Our Toolset

As we told colleagues and clients we were writing this book, they asked, "How can you write together? How can you know who does what? Do you have a board? How do you overcome the space-time continuum?" and "What tools do you use?" This appendix answers those questions.

Many distributed agile teams think they need to start with tools for the work. We started with communication tools.

Communication and Writing Tools

We used Zoom for our daily interaction. We used Google docs for our collaborative writing.

Zoom allowed us to see each other *as we wrote*. We could tell when the other person was excited or confused.

Google docs allowed us to pair-write. We could each write in the same document at the same time.

We used OmniGraffle for the images. We often shared the screen to collaborate on the images.

We also used iPhone Messaging when we couldn't talk to each other. We sometimes used Messaging to check in on each other. If we needed to cancel or reschedule, Messaging allowed us to use the most rapid asynchronous communication to let each other know. This was our primary backchannel.

Some information wasn't time-sensitive. We used email for things we needed to let each other know about at some point.

We used our Dropbox folder to save research articles. In addition, we used other folders to hold presentations and other material not necessarily for the *book*, but around the book product.

We also used a Todos file in Google docs, first organized by chapter. As we finished the tasks, we crossed them off.

Book Generation Tools

We wrote the book using Leanpub.com to publish. That meant we needed our content in Markdown at some point.

We tried—for about 10 minutes—to write directly into a Markdown editor in Google docs. Once we realized we couldn't both write together, we abandoned that attempt.

We next started writing using Google docs without any markup language. We tried a variety of export tools and plugins to try to make the Markdown "right." They didn't work that well for our publishing on Leanpub.

We decided to write in Markdown, in Google docs. We could then finish a file—whatever passed for finished—sweep it into Markdown and create a preview on Leanpub.

We couldn't finish an entire chapter in one writing session. Some chapters took several weeks. Some only took one week. And, because agile approaches are a system, we needed to refer back and forth in the book.

We used Johanna's notation of "XX:". The XX means "this isn't done; resolve later." Using that notation allowed us to continue to progress every day. When we wrote enough, we could search for the XXs and resolve them.

Integrating Reviewer Feedback

Every book writer has trouble integrating reviewer feedback. Reviewers offer feedback in a variety of forms. We needed to organize the feedback and integrate the feedback we chose to use.

We wanted to experiment with a different collaboration tool to see if it made the process easier. We tried Dropbox Paper to organize the initial reviewer comments.

Each chapter's comments became a new Dropbox Paper file. We could easily cut-and-paste the comments into the appropriate files.

Then, we could identify the subsections within a chapter and move the reviewer comments to the subsection part of the review notes in the Paper file. Dropbox Paper allows you to drag paragraphs, which made this sorting of reviewer comments faster.

We then wanted a way to signal when we had addressed a reviewer comment. Converting each reviewer comment into a Dropbox Paper task allowed us to "check off" the comment when we addressed it. This gave us a quick visual of progress in our responses to reviewer comments within a chapter.

We found this approach helpful for our first round of review. The first round of review comments helped us reshape the book. This is common with a non-fiction book. The writer rarely organizes correctly at the start of a project.

In the final round of reviews, we simply pulled chapter-specific errors and recommendations into one large Google doc file with each chapter as a new heading. As we completed each chapter, we crossed off the issue.

Tools We Didn't Use

We never used a board to visualize our progress. The book frame told us what we needed to do next.

We tried to write in chapter order. We didn't worry when we reorganized the chapters. We kept writing, staying in flow.

How We Lived the Mindset

We used these principles: be prepared to experiment with everything; everyone on the team needed equal access and training for our tools; and beware of change fatigue from too many changes at once.

We didn't experiment with everything at once. We tried something, lived with it for a while and, based on measurements or results, decided what to do.

We made sure to include each other on how to create graphics, how to use our book creation tools, and how to compile the book. We didn't pair on *everything* if one of us was better at a certain kind of work and the other was on the verge of change fatigue. For example, Mark understood the ideas of layers in images while Johanna was learning about them. He drew images and showed Johanna later. Johanna understood how to sweep the Google docs into Markdown and compile the book. She showed him so he could do it later.

We didn't "protect" each other from work, even if we experimented and temporarily confused each other. We managed our experimentation, toolset, and change fatigue to keep progressing on our product.

Would We Do It Again

We found the fewest tools and keeping a regular rhythm to our sessions were most important. There was never a time that we wrote solo. Our paired approach helped us feel joint ownership and commitment to the book.

For images, we often had one person create a first version in OmniGraffle and then we reviewed it together. Either of us could then edit the image later.

We divided some of the asynchronous work, such as collecting reviewer comments and other book-related administrative tasks. Those kinds of decisions did not require much interpretation or decision. We always created as a pair.

Before we ended each working session, we would agree to what we would review or take on as administrative tasks until the next time.

We both agree that we would likely work this way (and with each other) on future projects because we always collaborated on the creation process, experiments, and decisions. We also enjoyed making each other laugh.

Compass Activity for Distributed Teams

How often have you heard the following statements from team members?

- He just wants to jump in without thinking through the consequences.
- She takes too much time checking the facts.
- I'm tired of all his questions. Can we just get this done?
- Why do we have to "checkin" with everyone on how they feel about the project? It feels like therapy, not work.

If these statements sound familiar to you, your team might need to understand their collaboration preferences.

This activity helps team members build empathy for team mates by quickly learning individual preferences. This empathy building can be leveraged to develop working agreements for better collaboration within the team.

The activity runs for 45 to 60 minutes or can be expanded to deepen the understanding. It can also be used as part of other team meetings such as chartering or an early retrospective.

Prepare for the Compass Activity

Step 1: Prepare the Compass Diagram

Before bringing your team together, prepare a diagram that the team can interact with online (like the image below).

Select avatars for each member of the team. Some team members prefer to represent themselves as their profile image. Some prefer a different image, such as a beer mug for the person who likes craft beers, a purple pen for people who prefer purple, or a bicycle for people who like to cycle. You might ask the team what they prefer for their avatar for this activity.

Once the team selects the avatars they prefer, the team will be able to recognize who is where.

We've used generic icons in the images here.

Make sure all team members have edit access to the online document.

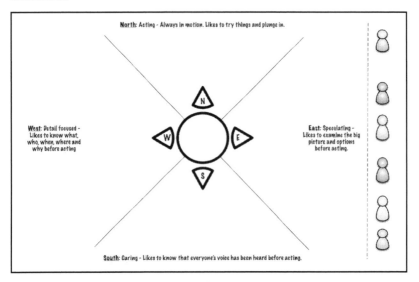

Figure B.1: Prepare the Compass Diagram

Step 2: Plan Breakout Sessions for Psychological Safety of the Team

The number of people in the team determine how you facilitate the activity. If you have a team of six or more people, consider breaking into pairs or triads. If you have a team of *fewer* than six people, consider an all-team discussion.

The point of the breakouts is to *discuss* the team's preferences' and differences in a psychologically safe environment.

A successful all-team discussion depends on the current psychological safety of the team. Have you seen team members refrain from raising certain topics or ideas? Is there any existing conflict between team members? If you answer "yes" to either of these questions, still consider breakouts as pairs.

You, as the facilitator, will also find it easier to keep the breakouts to a similar size. You may need to adjust the time in the breakout and the reminder time if this occurs. Try to stick to 1-2 minutes per question per compass point to keep answers short.

Step 3: Rehearse Transitions for the Online Breakouts

If you have not facilitated online breakouts before, plan your facilitation:

- Ensure you have an dedicated chat backchannel where you can reach all participants easily and at the same time.
- Practice the mechanics and timing of people moving into separate meeting spaces.
- Practice calling people back to the main space via the dedicated backchannel.

There are a number of ways to support the breakouts with a backchannel:

1. If possible, use the existing team's chat channel as the backchannel.
2. Prepare separate meeting rooms. Be ready to assign specific rooms to team members.

Step 4: Prepare a Display of "Results"

Each group representing a compass point will share their answers to a set of questions. Create multiple group documents, where the breakout group can answer these questions. In addition, create a team document where the entire team can synthesize the *team's* answers.

	North	East	South	West
Strengths?				
Limitations?				
Style most difficult to work with? Why?				
How to work effectively with my style?				
What is valued in the other styles?	East: South: West:	North: South: West:	North: East: West:	North: East: South:

Figure B.2: Table to Debrief Compass Preferences

Facilitate the Compass Activity

In this basic activity, you will take roughly 60 minutes for the following steps. Allow 2-3 minutes extra for any "adjustments." The steps to facilitate the activity are as follows:

Opening (2 minutes)

Welcome all team members to the activity and invite them to turn on their cameras for the activity as it will help them better understand how they each react during the activity. Explain the activity:

"This activity will help you and your fellow team members quickly discover your individual preferences. Once you have discovered and understood these preferences, we can use the results to develop working agreements for better collaboration within the team.

First we will have everyone individually indicate their preference, we will then break into smaller groups to discuss what we observe in those preferences, and then we'll come back together to share our observations. I will provide specific instructions at each step. Is everyone ready?"

Gather Data (5-7 minutes)

"Please review the compass diagram on the screen. Notice the four points of the compass are labeled:

- North: Acting—Always in motion. Likes to try things and plunge in.
- East: Speculating—Likes to examine the big picture and options before acting.
- South: Caring—Likes to know that everyone's voice has been heard before acting.
- West: Detail-focused—Likes to know what, who, when, where, and why before acting."

Provide these instructions:

"Consider the four compass points and then move your avatar to the compass point that most accurately captures how you typically work with others. You might feel like you fall between two points, but choose the one you feel most strongly aligned with. Any questions?"

Wait for a few seconds to make sure everyone understands.

"Then please move your avatar now."

Analyze Data in Breakouts (10-12 minutes)

Typically we move the team members into breakouts determined by the compass point they selected. You may provide instructions as follows:

"We will now go to our breakouts *based on our compass points* to answer the following questions:

- What are the strengths of your style? (in 3-4 words)
- What are the limitations of your style? (in 3-4 words)
- Which style do you find most challenging to work with and why?
- What do people from other styles need to know about you to work with you effectively?
- What do you value about the other 3 styles? (consider 1 word per style)

You only have <u>10</u> minutes and I will let you know when you have <u>5</u> minutes left.

Our breakouts are arranged as follows: (provide details for people to move to breakouts). Any questions?"

If you only have one person on a particular compass point, you may need to either:

- Join them to answer the questions in the breakout or
- Ask them to join another small breakout.
- If your team is *fewer* than six people, consider having an all-team discussion.

For teams of six or more people, consider breaking into pairs or triads, as long as the people feel safe to speak.

Don't worry if you don't see too many differences.

Be aware of teams that group around a compass point. While teams might not be homogeneous, teams may clump near two compass points. We've seen teams who clump around one compass point. For example, Operations Teams tend to be action-oriented. The action draws them to that kind of work.

Also consider how many people might you see gather at the various compass points. If you have a seven-person team, you might only see two or three compass points with avatars. Or, you might see 1-2 people at each compass point and one person straddling two points. Again, consider pairing people up to support psychological safety.

Debrief Together (22-32 minutes)

Ask each compass point breakout group to choose their spokesperson to share their results. Give them 2-3 minutes to summarize.

Then, ask for a volunteer group to debrief first (it may be North). Then move around the compass points asking each group to provide their answers to the questions in 4-6 minutes each. Scribe their responses as described in preparation Step 3 in an observations document.

Finally, ask for any key insights from the whole group (another 4-5 minutes) and record these in the observations document.

Consider These Other Facilitation Tips

We have found it useful to have the following tips in mind when facilitating this activity. You may need to provide some of these tips to the team members if questions arise.

- You may alter the debrief to include time for the team to brainstorm and vote on working agreements. You might allow another 30 minutes or hold a second session.
- Even if you suggest that team members place their avatar in the one quadrant that represents where they most often tend to act, some may feel strongly about placing themselves on the border between two preferences (see example below). This is perfectly fine and you might explore this in your debrief with the entire team with some of the following statements and questions:
  - How might these combined preferences help you collaborate as team members?
  - How might they be a challenge?
  - What might need to be considered if the two sets of preferences appear to be competing?

Figure B.3: Avatars Moved in the Compass Activity

- Notice if someone places their avatar in the center of the diagram. Too often, this means the person doesn't feel safe. We do not recommend you address this in the meeting. Instead, review the ideas in *Create Psychological Safety in Your Team* on page 88 and follow up later with a 1-on-1 conversation.

- When adding a new member to the team, re-run the activity with a clean slate. Team members may want to use or refer to the old compass diagram to "save time." Let the team know that this might put the new team member at a disadvantage as they may feel they need to "study" the old diagram instead of using this time in the activity to discover each other's preferences. More senior members may notice once they run the activity again that they have forgotten the nuances of some of the team mates' preferences or there might even be a slight shift in preferences due to the collaboration on the team. This makes re-running the activity as a new activity even more valuable for the team.

- Guests and non-team members? As you begin to use this activity with multiple teams in your organization, guests may wish to observe the activity. Politely discourage this, explaining that guests may impact the psychological safety and openness of the team members. However, some teams may have non-team members join them who occasionally collaborate with the team, based on their subject matter expertise or skills. These non-team member participants can be useful if they interact with the team frequently enough and are willing to abide by the preferences and working agreements of the team. In such situations, it's always best to ask the team first how they may feel about having someone outside the team attend such an event. Be sure to explain that everyone will be discussing their preferences and the non-team member may or may not be participating.
- The original activity described in the references mentions 20 minutes for the basic activity. Typically, a distributed team needs more time for these kinds of conversations as they may not be accustomed to them.

If you want to expand this activity for deeper learning, please see the references.

References

Adapted from:

- http://schoolreforminitiative.org/doc/compass_points.pdf
- https://www.kqed.org/mindshift/40880/a-simple-exercise-to-strengthen-emotional-intelligence-in-teams

Annotated Bibliography

[AMA11] Amabile, Teresa and Steven Kramer. *The Progress Principle: Using Small Wins to Ignite Joy, Engagement, and Creativity at Work.* Harvard Business Review Press, Boston, 2011. They have completed the research that says we like to finish work in small chunks so we can make progress.

[BRO95] Brooks, Frederick P. *The Mythical Man-Month: Essays on Software Engineering, Anniversary Edition (2nd Edition)* Addison-Wesley, Boston, 1995. How software projects and teams really work. Learn from a master.

[COH05] Cohn, Mike. *Agile Estimating and Planning,* Prentice Hall, New Jersey, 2005. The seminal work on estimating and planning for agile projects.

[COC06] Cockburn Alistair. *Agile Software Development: The Cooperative Game, 2nd ed.* Addison-Wesley, Boston, 2006. One of the first books describing agile principles and practices and defining the importance of face-to-face communication in the cooperative game of software.

[COV89] Covey, Stephen R. *The 7 Habits of Highly Effective People: Powerful Lessons in Personal Change.* Simon and Schuster, 1989. We have found that these habits help us structure our lives and our leadership for the better.

[CSI08] Csikszentmihalyi, Mihaly. *Flow: The Psychology of Optimal Experience*. HarperCollins Publishers, 2008. Learn how people finish work, separately and together.

[DAL86] Daft, Richard L. and Robert H. Lengel. *Organizational Information Requirements, Media Richness and Structural Design* in *Management Science*, Vol. 32, No. 5, Organization Design. (May, 1986), pp. 554-571. The oft-cited article about media richness and how the media can help or hinder resolution of uncertainty and equivocality.

[DER06] Derby, Esther and Diana Larsen. *Agile Retrospectives: Making Good Teams Great*. Pragmatic Bookshelf, Dallas, TX and Raleigh, NC, 2006. The classic work about agile retrospectives.

[EOH13] Eoyang, Glenda H. and Royce J. Holladay. *Adaptive Action: Leveraging Uncertainty in Your Organization*. Stanford Business Books, 2013. Teams are complex adaptive systems. The first book to describe how containers, differences, and exchanges influence the entire adaptive system.

[EDM12] Edmondson, Amy C. *Teaming: How Organizations Learn, Innovate, and Compete in the Knowledge Economy*. Jossey-Bass, San Francisco, 2012. How self-organized teams really work, and what we need to make them work in different cultures.

[ERR11] Carmel, Erran & J. Alberto Espinosa. *I'm Working While They're Sleeping: Time Zone Separation Challenges and Solutions*. Nedder Stream Press, 2011. A terrific general-purpose read about geographically distributed people and approaches you might use to manage the time zone differences.

[ERR99] Carmel, Erran. *Global Software Teams: Collaborating across Borders and Time Zones*. Prentice Hall. 1999. The first book that discusses the ideas of how global teams might work.

[KAN14] Kaner, Sam *Facilitator's Guide to Participatory Decision-Making, 3rd Edition*. Jossey-Bass. 2014. If you need to facilitate any meetings at all, use the ideas in this book. We both find it an invaluable resource.

[KAT99] Katzenbach, Jon R. and Douglas K. Smith. *The Wisdom of Teams: Creating the High-Performance Organization.* HarperCollins Publishers, New York, 1999. Another classic, about differentiating workgroups from teams and what creates a team.

[KOC05] Kock, Ned. *Media Richness or Media Naturalness?* in IEEE Transactions on Professional Communication, Vol 48, No. 2, June 2005, pp. 117-130. Excellent rebuttal to communication richness.

[MOA13] Modig, Niklas and Pär Åhlström. *This is Lean: Resolving the Efficiency Paradox.* Rheologica Publishing, 2013. Possibly the best book about how managers should consider agile and lean. A wonderful discussion of resource efficiency vs. flow efficiency.

[POP03] Poppendieck, Mary and Tom Poppendieck. *Lean Software Development: An Agile Toolkit.* Addison-Wesley, Boston, 2003. The first book to provide a lean approach to software. Contains the simplified value stream mapping we show here.

[BCD05] Rothman, Johanna and Esther Derby. *Behind Closed Doors: Secrets of Great Management.* Pragmatic Bookshelf, Dallas, TX and Raleigh, NC, 2005. We describe the Rule of Three and many other management approaches and techniques in here.

[ROT07] Rothman, Johanna. *Manage It! Your Guide to Modern, Pragmatic Project Management.* Pragmatic Bookshelf, Dallas, TX and Raleigh, NC, 2007. Waterfall and agile are just two of the four major kinds of life cycles. You can select which life cycle will make the most sense for you. If you want to examine other life cycles and learn how to replan, read this book.

[RE14] Rothman, Johanna and Jutta Eckstein. *Diving for Hidden Treasures: Uncovering the Cost of Delay in Your Project Portfolio.* Practical Ink, 2014. A book about Cost of Delay and how to see how those costs affect your project portfolio.

[RHI13] Rothman, Johanna and Shane Hastie. *Lessons Learned from Leading Workshops about Geographically Distributed Agile Teams* in *IEEE Software*, March/April 2013, pp 7-10. Shane and Johanna developed and facilitated several workshops about geographically distributed teams in a variety of locations between 2011-2014. This article captures our lessons learned from the workshops. The article is available at https://www.jrothman.com/articles/2013/03/lessons-learned-from-leading-workshops-about-geographically-distributed-agile-teams/.

[ROT12] Rothman, Johanna. *Hiring Geeks That Fit.* Practical Ink, 2012. If you want to know how to hire people, this is it, from soup to nuts. All the templates are available for free on Johanna's website. The book explains how to use them.

[ROT16A] Rothman, Johanna. *Agile and Lean Program Management: Scaling Collaboration Across the Organization.* Practical Ink, 2016. A program is a collection of projects with one business objective, often requiring several feature teams. Learn how to scale the collaboration, not the process.

[ROT17] Rothman, Johanna. *Create Your Successful Agile Project: Collaborate, Measure, Estimate, Deliver.* Pragmatic Bookshelf, Raleigh, NC, 2009. Learn the difference between iteration- and flow-based agile approaches and what practices to consider when. Includes how-tos for workgroups and managers.

[ROR98] Smith, Preston and Donald J. Reinertsen. *Developing Products in Half the Time: New Rules, New Tools.* Wiley, 1998. A classic introduction to many ideas of lean and agile for product development.

[SCH10] Schein, Edgar H. *Organizational Culture and Leadership.* Jossey-Bass. San Francisco 2010. Culture is not about the color of the walls or the foosball tables. Culture is about us, as humans. A fascinating look at what culture means.

[SUN18] Sutherland, Lisette and Kirsten Janene-Nelson. *Work Together Anywhere: A Handbook on Working Remotely—Successfully—for Individuals, Teams, and Managers.* Collaboration Superpowers, Delft, the Netherlands. 2018. Full of tips and ideas for managers and team members about how to facilitate remote work. Includes guidance on preparing for a transition to remote work for any kind of team.

[TAB06] Tabaka, Jean. *Collaboration Explained: Facilitation Skills for Software Project Leaders.* Addison-Wesley, 2006. One of the first books to introduce skills for planning and facilitating various agile meetings such as workshops, sprint planning, release planning, retrospectives, and more.

[WEI93] Weinberg, *Gerald M. Software Quality Management, Vol 2: First-Order Measurement.* Dorset House Publishing, New York, 1993. This particular Weinberg book has excellent definitions of congruence and human interaction.

Glossary

Agile Approach: A collaborative team-based approach to finishing valuable work. The value of working in an agile way is that you have the ability to replan quickly, because the team completes work.

Backchannel: An application that allows everyone on the team to communicate and coordinate when another communication channel may be in use (audio/video). We particularly like a chat application for a team's backchannel.

Backlog: Ranked list of items that need to be completed for the product.

Bus factor: The risk of not sharing information among all team members. When a team has a bus factor of "one," that means only one team member has the necessary information.

Buddy System: People who help a remote team member maintain connection for the work.

Collocated: When team members sit side by side, or, at least when everyone is within 30 meters of each other.

Copilot: A person who assumes a similar role to assist across multiple groups or time zones. For instance, a Product Owner or a meeting facilitator in one location may have a copilot in another location who can help coordinate activities and responsibilities.

Community of Practice: A way to share knowledge among people who belong to different teams, and share the same interests or function. For example, in a program, you might have an architecture community of practice that helps any developer learn how to evolve the design of the product. A test community of practice would provide a forum for testers to discuss what and how to test.

Conway's Law: The architecture of your product reflects the location/architecture of the team who created that product. Originally described in *The Mythical Man Month*.

Cost of Delay: The revenue impact you incur when you delay a project. Aside from "missing" a desired release date, you can incur Cost of Delay with multitasking, or waiting for experts, or from one team waiting for another in the program. All of these problems—and more—lead to delay of your product release.

Digital nomad: Knowledge worker who works from anywhere as long as they have electricity and a high-speed network connection. Nomads include anyone who might travel to a new location (or country) every 2-3 months for a change of scenery and to optimize their preferred work environment.

Distributed: When some team members are separate from each other, the team is distributed.

Dispersed: When none of the team members share any collocated space, the team is dispersed.

Double Loop Learning: Assess and challenge everything about the product and the project. Because agile approaches allow the team to finish frequently, the team (and the Product Owner or customer) can reassess their product progress and assumptions, as well as the project and process assumptions.

Flow: Instead of planning for a timebox of a week or two, the team limits the number of items under consideration. The team still takes the work in rank order.

Generalizing Specialist: Someone who has one skill in depth, and is flexible enough to be able to work across the team to help move a feature to done.

Geo-fence: The teams or team members are separated by more than 30 meters, even if they are all in one zip code.

HiPPO: Highest Paid Person's Opinion. A HiPPO can derail any decision about anything.

Information Radiator: Any physical or virtual board that provides up-to-date information on the status of the work or the product.

Kanban: Literally the Japanese word for "signboard." A scheduling system for limiting the amount of work in progress at any one time.

Kaizen: A philosophy for continuous improvement.

Lean: A pull approach to managing work that looks for waste in the system, a holistic approach. The two pillars of lean are respect for the people and continuous improvement.

Lean Coffee: A meeting structure where the people attending the meeting evolve the agenda as they proceed, always finishing a conversation on the most valuable item first. See http://leancoffee.org/.

MVP: Minimum viable product. What is the minimum you can do, to create an acceptable product? This is not barely good enough quality. This is shippable product. However, this is minimal in terms of features.

Pairing: When two people work together on one task, one monitor, and one keyboard. They work only on this task together.

Parking Lot: This is a place to put issues you don't want to lose but don't necessarily want to address at this time.

Psychological Safety: Each team member has the capability and motivation to say what they perceive and feel, in an environment of respect for each other.

Spike: If you cannot estimate a story, timebox some amount of work (preferably with the entire team) to learn about it. Then you will be able to know what to do after the day or two timebox.

Servant Leadership: An approach to managing and leading where the leader creates an environment in which people can do their best work. The leader doesn't control the work; the team does. The leader trusts the team to provide the desired results.

Sprint: An iteration in Scrum.

Story or User Story: A requirement in the form of value to a person.

Swarming: When the team works together to move a feature to done, all together.

Technical Debt: Shortcuts a team takes to meet a deliverable. Teams might incur technical debt on purpose, as a tactical decision. Technical teams can have architectural, design, coding, and/or testing debt. Program teams might have risk or decision debt—the insufficiency of work for managing risks or making decisions.

Timebox: A specific amount of time in which the person will attempt to accomplish a specific task.

WIP or Work in Progress: Any work that is not complete. When you think in lean terms, it is waste in the system. Note that you do not get credit for partially completed work in agile approaches.

Index

More from Johanna

I consult, speak, and train about all aspects of managing product development. I provide frank advice for your tough problems—often with a little humor.

If you liked this book, you might also like the other books I've written:

Create Your Successful Agile Project: Collaborate, Measure, Estimate, Deliver (https://www.jrothman.com/cysap)

Manage Your Project Portfolio: Increase Your Capacity and Finish More Projects, 2nd ed (https://www.jrothman.com/MYPP)

Agile and Lean Program Management: Scaling Collaboration Across the Organization (https://www.jrothman.com/ALPM)

Diving for Hidden Treasures: Uncovering the Cost of Delay Your Project Portfolio (https://www.jrothman.com/diving)

Predicting the Unpredictable: Pragmatic Approaches to Estimating Project Cost or Schedule (https://www.jrothman.com/predict)

Project Portfolio Tips: Twelve Ideas for Focusing on the Work You Need to Start & Finish (https://www.jrothman.com/port-tips)

Manage Your Job Search (https://www.jrothman.com/myjs)

Hiring Geeks That Fit (https://www.jrothman.com/hiring)

Manage It!: Your Guide to Modern, Pragmatic Project Management (https://www.jrothman.com/manageit)

Behind Closed Doors: Secrets of Great Management (https://www.jrothman.com/bcd)

In addition, I have essays in:

Readings for Problem-Solving Leadership (https://leanpub.com/pslreader)

Center Enter Turn Sustain: Essays on Change Artistry (https://leanpub.com/changeartistry)

I'd like to stay in touch with you. If you don't already subscribe, please sign up for my email newsletter, the Pragmatic Manager (https://www.jrothman.com/pragmaticmanager/), on my website. Please do invite me to connect with you on LinkedIn (https://www.linkedin.com/in/johannarothman), or follow me on Twitter, @johannarothman.

I would love to know what you think of this book. If you write a review of it somewhere, please let me know. Thanks!

—Johanna

More from Mark

With over two decades of experience in agile principles and practices, Mark Kilby has cultivated more distributed and dispersed teams than collocated teams. He has consulted with organizations across many industries and coached teams, leaders and organizations internally. Mark also co-founded a number of professional learning organizations such as Agile Orlando, Agile Florida, Virtual Team Talk, and the Agile Alliance Community Group Support Initiative among others. His easy-going style helps teams learn to collaborate and discover their path to success and sustainability. Mark shares his insights on distributed and agile teams in dozens of articles in multiple publications. Most of his latest ideas and developments can be found on www.markkilby.com.

We practice what we preach.

We both offer or will offer virtual workshops.

Johanna's online workshops page:
https://www.jrothman.com/services/online-workshops/.
She offers a variety of workshops, including writing
workshops, with more planned. She announces
new workshops on her newsletter, the
Pragmatic Manager, and on her blog.
See https://www.jrothman.com for all the details.

Mark's current workshops on leading successful
distributed agile teams are available at
https://www.markkilby.com/workshops/.
He'll announce new workshops via his newsletter at
https://www.markkilby.com.

20478152R10179

Printed in Great Britain
by Amazon